Measurement of Gender Empowerment Equity and Development

The Authors

Dr. C. Satapathy served Odisha University of Agriculture and Technology with different capacities and retired as Professor and Dean of Extension. Presently he is working as Director of Amity, Bhubaneswar, Odisha. To his credit, he has guided many Ph.D. scholars, published sixteen books and many research papers in national and international journals. He was awarded Gold Medal in M.Sc. (Ag.), Extension level and obtained Ph.D. degree from IARI, New Delhi in 1981.

Dr. Sabita Mishra is presently working as Senior Scientist (Agriculture Extension) at the Directorate of Research on Women in Agriculture, ICAR, Bhubaneswar. She obtained her Ph.D. Degree from Utkal University in 1991. To her credit, she worked in the field of rural extension besides publishing eight important books, many research papers in national and international journals. She has been actively involved in research for Women in Agriculture, organizing capacity building programmes for Agricultural Scientists and field programmes over last nineteen years. She was awarded by His Excellency, Governor of Odisha for her immense contribution in the field of empowerment of rural women.

Measurement of Gender Empowerment Equity and Development

C. SATAPATHY
Director
Amity Global Business School
Bhubaneswar, Odisha

SABITA MISHRA
Senior Scientist (Agriculture Extension)
Directorate of Research on Women in Agriculture
Bhubaneswar, Odisha

2015
Daya Publishing House®
A Division of
Astral International Pvt. Ltd.
New Delhi – 110 002

Published by : **Daya Publishing House®**
A Division of
Astral International Pvt. Ltd.
– ISO 9001:2008 Certified Company –
4760-61/23, Ansari Road, Darya Ganj
New Delhi-110 002
Ph. 011-43549197, 23278134
E-mail: info@astralint.com
Website: www.astralint.com

Laser Typesetting : **Classic Computer Services**, Delhi - 110 035

Printed at : **Replika Press Pvt. Ltd.**

PRINTED IN INDIA

Preface

The present concept of women empowerment is viewed as degree to which they participate and contribute to national development. It is necessary to measure quantitatively and qualitatively the level of empowerment, equality and development that we claim as outcome of our efforts. The book deals with different approaches to measure gender empowerment equity and development with scope for future improvement. The authors believe and hope that the scholars and research workers dealing with women empowerment will be benefitted by the book.

C. Satapathy

Sabita Mishra

Contents

Preface   *v*

List of Figures   *ix*

List of Tables   *xi*

1. **Status of Women**   **1**
   Parameters of Empowerment   3
   Women and Development   5

2. **Measurement of Gender Empowerment, Equity and Development**   **7**
   Measurement of Empowerment   8
   Social Variables   10
   Economic Variables   11
   Psychological Variables   14
   Development Variables   15
   Social Empowerment   23
   Economic Empowerment   24
   Political Empowerment   24
   Legal Empowerment   26
   Gender Development Index   26

3. **Measurement of Traits of Women Leaders**   **28**
   Trait Approach   29
   Skill Approach   32
   Style Approach   33
   Situational Approach   35

4. Measurement of Capacity Building and Knowledge Gain   **42**
Capacity Building   42
Training   43
Impact Evaluation of Training Programme   49

**5. Measurement of Adoption: Behaviour and Innovativeness
of Farm Women**   **56**

6. Measurement of Communication Behaviour of Farm Women   **65**
Women as Source of Communication   67
Indicators for Measurement   67
Credibility of Communicator   68
Measurement of Innovativeness of Rural Women   72

7. Measurement of Managerial Ability of Rural Women   **76**
Household Activities   76
Care of Children and their Education   78
Resource Management   80
Management of Social Functions and Community Activities   82

References   **87**

Index   **89**

List of Figures

Figure 1: Components of Empowerment   3

Figure 2: Ladder for Moving up Women   5

Figure 3: Indicators of Empowerment   9

Figure 4: Interrelationship between Variables   9

Figure 5: Dependent Variables for Gender Equity   27

Figure 6: Promoting Leadership Effectiveness   29

Figure 7: Trait Theory of Leadership Style   30

Figure 8: Components of Skill Model   32

Figure 9: Factors Influencing Groups and Leaders   39

Figure 10: Capacity Building   43

Figure 11: Experiential Learning   44

Figure 12: Training Programme   50

Figure 13: Teaching and Learning Process   50

Figure 14: Impact of Training Programme   51

Figure 15: Training Result   52

Figure 16: Communication Domain of Farm Women   66

Figure 17: Dimensions of Managerial Ability of Rural Women   77

Figure 18: Managerial Ability   82

Figure 19: Indicators of Competencies   84

Figure 20: Management Practices   85

Figure 21: The Livelihood Pentagon   85

Figure 22: Concept of Rural Livelihood System   86

List of Tables

Table 1: Women Representation in Parliament   2

Table 2: Variables and Indicators of Gender Empowerment   4

Table 3: Scoring Pattern   10

Table 4: Farmers Category   11

Table 5: Access and Control Over Resources   19

Table 6: Gender Equity Index   20

Table 7: Areas of Decision-making   21

Table 8: Traits of all Indicators   31

Table 9: Relative Position of Traits of Leadership   31

Table 10: Technical Skill   32

Table 11: Human Skill   33

Table 12: Conceptual skill   33

Table 13: Style Approach   34

Table 14: Situational Leadership Style   36

Table 15: Imagining Leaders by Men and Women   36

Table 16: World Scenario of Perceived Leadership   38

Table 17: Region-wise Cumulative Number of SHGs in India   39

Table 18: Region-wise SHG Federation in India   40

Table 19: Self Help Group in Odisha   40

Table 20: District-wise Women Self Help Groups   40

Table 21: Evaluation Frame Work   51

Table 22: Before - After Design   54

Table 23: Average Score in Pre and Post-Measurement   54

Table 24: Covert Behaviour and Overt Behaviour   57

Table 25: Adoption Stage 'NEED'   58

Table 26: Adoption Stage 'AWARENESS'   59

Table 27: Adoption Stage 'DELIBERATION'   59

Table 28: Adoption Stage 'INTEREST'   59

Table 29: Adoption Stage 'TRIAL'   59

Table 30: Adoption Stage 'EVALUATION'   60

Table 31: Adoption Stage 'ADOPTION'   60

Table 32: Adoption Stage 'INTEGRATION'   60

Table 33: Stages of Adoption Process   60

Table 34: Women in Adoption Behaviour   61

Table 35: Movement of Men and Women at different Stages of
          Adoption Process   61

Table 36: Classification of Adopters   62

Table 37: Measurement of Innovativeness of Individuals in
          Adopter Category   62

Table 38: Characteristics of Early Adopters   62

Table 39: Characteristics of Early Majority   63

Table 40: Characteristics of Late Majority   63

Table 41: Characteristics of Laggards   64

Table 42: Managerial Ability of Rural Women on Household Activities   77

Table 43: Managerial Ability of Rural Women on Child Care and Education 78

Table 44: Managerial Ability of Farm Women on Resource Management   80

Table 45: Managerial Traits of Rural Women   83

Table 46: Livelihood Concept and Components   86

Chapter 1

Status of Women

The Millennium Development Goal (MDG) recognizes the intrinsic and instrumental value of gender equity. The 2010 MDG summit concluded with the adoption of a global action plan to achieve the eight goals by 2015.The summit also adopted a resolution calling for action to ensure gender equity in education and health, economic opportunities and decision-making through gender mainstreaming by development of policy making. The resolution and action plan reflect the belief of the international development community that gender equality and women empowerment are development objectives in their own right (MDG 3 and 5) as well as this serving as critical channel for achieving the other MDGs and reducing income and non-income poverty. Gender equality and women's empowerment help to promote universal primary education (MDG 2), reduce under five mortality (MDG 4), improve maternal health (MDG 5) and reduce the likelihood of contracting HIV/AIDS/(MDG 6) (World Development Report 2012, p. 4).

Women in India and their issues and empowerment have gained importance since last two decades. The Government of India and public in general have realized the problems like dowry system, female foeticide, female infanticide, discrimination etc. The three important models have been examined to eliminate these problems. These are: (i) pluralism which argues that women should maintain and develop many of their distinctive characters while those circumstances which results in inequality; (ii) assimilation model which argues that women should try and shed their differences and join mainstream of the society; (iii) hybrid model proposes a society in which role of men and women are similar and need a change in traditional role of both sexes. These three models are in operation.

Current frame work of international development recognizes women empowerment as an immense effective strategy for all round development of the society. Though India is developing economically and technologically by leaps and bounds but women especially rural women still continue to be discriminated and their current status in the society still causes concern. The Government has attempted to involve and encourage rural women in decision-making process by providing one-third reservation for them in grampanchayat to ensure their participation at the local and district levels of governance through 73rd Amendment of the Constitution. At present there are 260,000 panchayats representatives in India out of which around 75,000 are women. This is the largest number of elected women in the world. Besides, women representation in parliament also is at increasing rate although not upto expectation.

Table 1: Women Representation in Parliament

Year	Total Number of Seats	No. of Women Members	Per cent to Total
1952	499	22	4.4
1957	500	27	5.4
1962	503	34	6.8
1967	523	31	5.9
1971	521	22	4.2
1977	544	19	3.5
1980	544	28	5.1
1984	517	44	8.1
1989	544	27	5.2
1991	544	39	7.2
1996	543	40	7.4
1998	543	43	7.9
1999	543	49	8.8
2004	543	45	8.2
2009	543	59	10.1
2014	543	61	11.23

Empowerment of women commonly refers to women. The target of our planning is to empower rural women as they form about half of the total population. India has been an agrarian country. Women constitute about 66 per cent of the agricultural workforce. Around 48 per cent self employed farmers are women and 64 per cent of the informal

According to UNICEF "State of World Children 2009" report 47 per cent of the India's women aged 20-24 were married before the prescribed legal age of 18 years, with 56 per cent in rural areas. About 40 per cent of world's child marriage occurs in India.

sector workforce depending on agriculture is women. Rural women provide food security to the country's 1.13 billion people. India is the home to 40 per cent of the world's underweight children and ranks 126 out of 177 countries in the UNDP Human Development Index. Annual average farmer suicide increased from 15,747

(1997-2001) to 17,366 (2002). But women despite their unbearable hardship and commitment to their children for food, health and education, have beyond doubt demonstrated their loyalty to financing banks through above 95 per cent repayment of loans.

A new approach to empowerment is visible through Self Help Group (SHG) units. It was only after mid 1990s the most rural women slowly and steadily found opportunity to access to credit through efforts of SHG. SHG linked credit facilities has covered 3.47 million SHGs and 45.1 million households. More than 50 per cent of the SHGs comprise women borrowers.

Parameters of Empowerment

Women empowerment is a term which indicates status of women in society compared to men in all spheres of development. Women comprise 50 per cent of the population, contributes 75 per cent work hours and receives 10 per cent income and 1 per cent share in property (FAO). The empowerment implies participation of women in decisions that affect their lives and enable them to build their strength and assets. The economic assets of both men and women is meager. Their social, political, environmental and personal assets assume much important for them. Building the assets of the poor and empowering them is thus the starting point for eradication of poverty. The strategy needs: (i) policy reforms that would enable to have access to assets; (ii) to ensure education and health to all and (iii) social safety.

Gender equity is essential for empowering women and men for eradication of poverty. The term empowerment is thus can be operationalized as: (i) removal of all discrimination against girls starting from birth in all aspects; (ii) empowering women

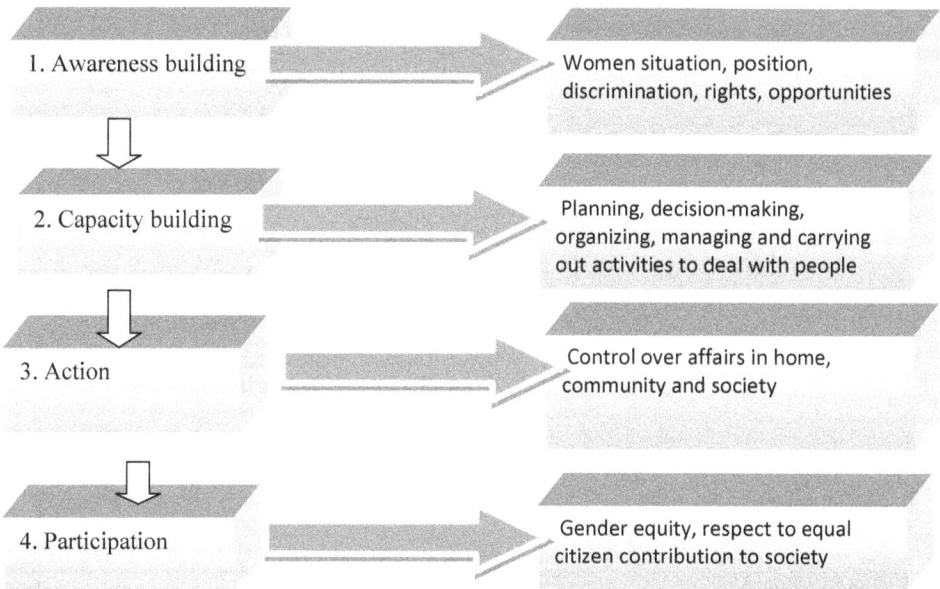

1. Awareness building	Women situation, position, discrimination, rights, opportunities
2. Capacity building	Planning, decision-making, organizing, managing and carrying out activities to deal with people
3. Action	Control over affairs in home, community and society
4. Participation	Gender equity, respect to equal citizen contribution to society

Figure 1: Components of Empowerment.

by giving them equal rights and access to land, credit and job opportunity and (iii) taking appropriate action for ending all forms of violence against women.

Empowerment is a process–not a thing which cannot be given to women as such. The process of empowerment is both individual and collective since it is through self and then involvement in groups that women begin to develop their awareness and ability to organize to take action and initiate change. Women empowerment can be considered as continuum of several inter-related, interwoven and interactive mutually reinforcing components.

In empirical studies for assessment of empowerment of women, there is need to select independent variables in relation to empowerment of rural women. A number of studies have been conducted to assess the level and type of empowerment. The measuring variables are of four types, namely: (i) socio-personal psychological variables, (ii) gender equity index, (iii) gender empowerment index and (iv) gender development index (UNO). The variables for measuring empowerment are:

Table 2: Variables and Indicators of Gender Empowerment

Variables	Indicators
1. Socio-personal and psychological variables	1. Farm size
	2. Educational status
	3. Marital status
	4. Family size
	5. Economic status
	6. Discrimination against girl child
	7. Family income
	8. Economic productivity capacity
	9. Managerial abilities
	10. Income generating
	11. Social participation
	12. Social mobilization
	13. Self orientation
	14. Extension participation
	15. Mass media exposure
	16. Perception of drudgery
	17. Access to basic facilities
2. Gender equity index	1. Education
	2. Employment
	3. Health and Nutrition
	4. Time
	5. Access and control over resources
3. Gender empowerment	1. Decision-making autonomy
	2. Social/personal autonomy
	3. Economic autonomy
	4. Political autonomy
	5. Legal autonomy
4. Gender development index	1. Life expectancy
	2. Education
	3. Income

According to the World Bank (2012) the main factors for moving up ladder by women are:

1. Occupation and economic change
2. Financial management
3. Education and training
4. Social network
5. Individual behaviour
6. Family status
7. Psychological traits
8. Marital status
9. Leadership

Women and Development

The present concept in development is the inclusion of women who represent about half of our population. The core issue of development may be of any field revolve around the participation of women. The main message of World Development Report 2012 highlight greater gender equality can bring development in sustainable manner benefitting the next generations. The major emphasis of development in

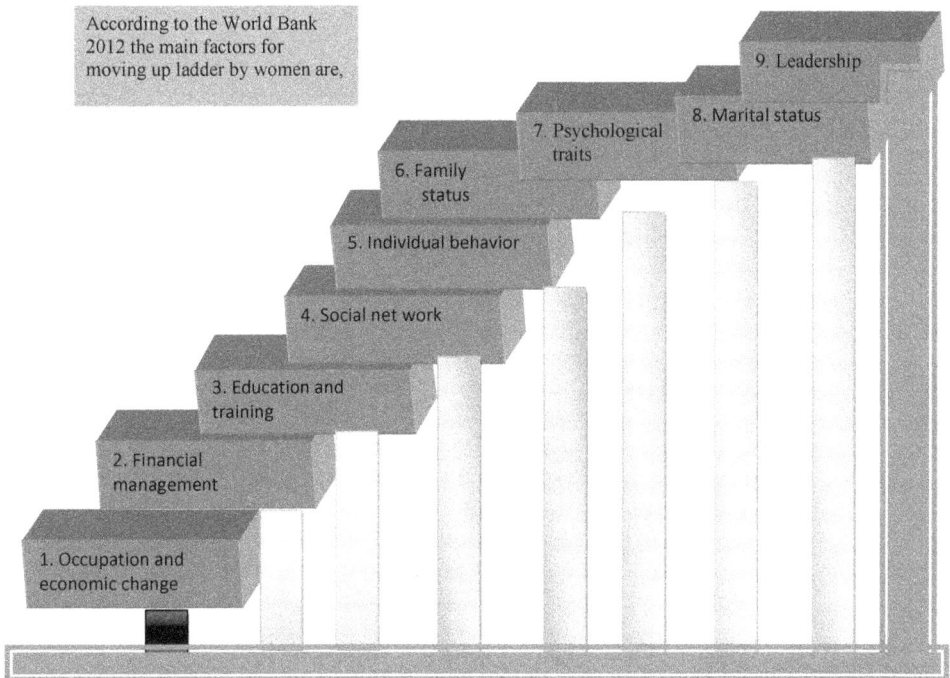

According to the World Bank 2012 the main factors for moving up ladder by women are,

9. Leadership
8. Marital status
7. Psychological traits
6. Family status
5. Individual behavior
4. Social net work
3. Education and training
2. Financial management
1. Occupation and economic change

Figure 2: Ladder for Moving up Women

context of gender equality is: (i) productivity gain: (ii) improved outcomes for next generation and (iii) increased presentation in decision-making process.

The achievement of gender directed programs are:

1. Educational enrolment of girl child
2. Increased life expectancy of women
3. Employment participation.

Spectacular achievements are observed in above three dimensions all over world specially developing and under developed nations. In India these dimensions have assumed greater responsibility and also response.

The report indicated considerable gap persists in the following areas of gender equality:

1. Excess mortality rate of women and girl child
2. Disparities in school of gild child
3. Unequal access to economic opportunities
4. Voice in decision-making process
5. Market, institutions are beyond the reach of many

Visible progress	Persistent gaps
1. Life expectancy	1. Health and sanitation
2. Girl education	2. Women and girl child mortality
3. Reduction in children	3. Economic opportunities
4. Women market network	4. Gap in earning capacity
5. Creation of women friendly organization	5. Low voice in decision making
6. Employment opportunity	6. Gender segregation around economic opportunities
7. Decline in fertility	

Chapter 2

Measurement of Gender Empowerment, Equity and Development

Empowerment can be achieved at six important levels. These are: (i) Economic front; (ii) Socio-cultural level; (iii) Family or interpersonal level; (iv) Legal level; (v) Political level and (vi) Psychological level.

1. Economic Front

The empowerment at family level is the scope for self employment, involvement in income generating activities, contribution to family income and control over family resources. At society level, access to credit facilities, employment and ownership over land, association and involvement in micro enterprises lead women to be empowered. At state and national level empowerment can be achieved through budget provision for their development and representation in policy decisions.

2. Socio-cultural Level

The empowerment of women at socio-cultural level is relatively more important in view of recognition and status. At family, the freedom for movement, non-discrimination and equal priority for education of girls reveal the real means and ways to empowerment. At society or community level their participation in community activities and women association are the indicators of empowerment.

> Reservation of seats for women is no doubt an innovative and important step to empowerment but their capacity building is much more important to perform role and responsibility.

Women representation in media world high lighting their roles and achievement reveal the sense of empowerment.

3. Family and Interpersonal Level

Empowerment of women at family sphere can be achieved in terms of involvement in decision-making process in both economic and social issues, control over sexual relation, ability to make decision about child bearing and free of domestic violence. At society level, they are to be empowered in the matters of autonomy, self selection of spouse, removal of dowry and campaign against domestic violence. At state or national level provision and policy of health care and education can bring empowerment among the rural women.

4. Legal Level

At family or individual level women need to be having knowledge about women's right, family support and encouragement to exercise their rights. At society level right to enjoy equality with men and enforcement of legal rights enhances empowerment of women. At state or national level, strict enforcement of legislation protecting women right would promote empowerment. Media support is a must for ensuring and implementation of legislations relating to women.

5. Political Level

Empowerment of women can be achieved in political level. The women are to know the political system and democratic ways of putting forward the problem before authorities. At society level their representation in local bodies and campaign are the indicators of empowerment. Women representation in state and national government like state assembly and parliament gives indication of their empowerment.

6. Psychological Level

Respect for women at family and community level reveals their empowerment. At state and national level their voices are to be respected and reflected in decision-making process if they are to be empowered. Women reservation bill has brought excellent result. A record of 59 MPs have been elected to the 15th parliament the highest ever since independence and 17 of them are aged less than 40 years. A majority of 23 MPs are from Congress. The BJP has 13 women members. But presence of women in Parliament or State Assembly is not enough unless they exercise their potentiality in nation building process. Empowerment includes encouraging and developing the skill for self-sufficiency. The ten important indicators of empowerment of women have been found out by the researchers as presented herewith.

Measurement of Empowerment

Measurement of empowerment can be made after deciding independent and dependent variables. Independent variables are those which can produce effect on empowerment. The interrelationship between these two can be stretched as far as possible to determine effect of each variable.

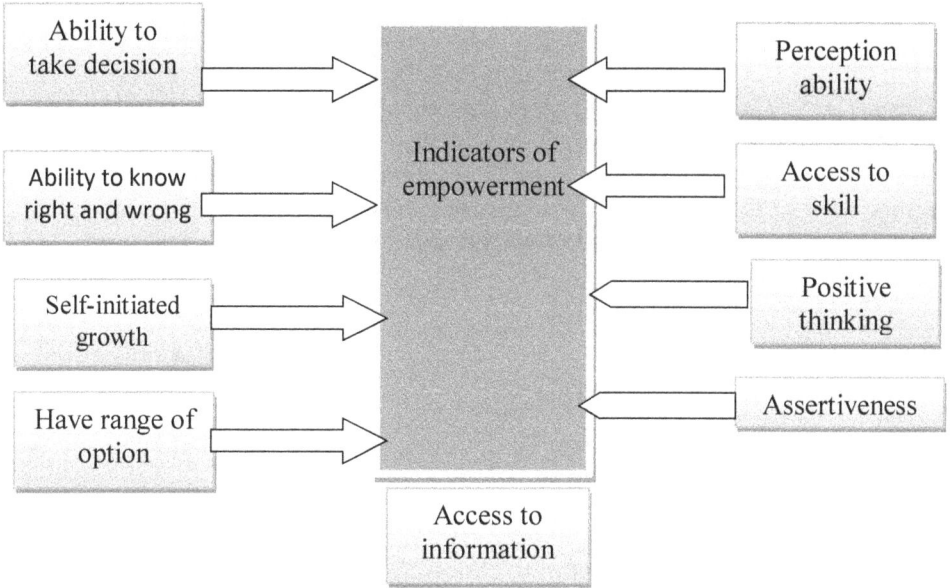

(Concept: Women at work, Satapathy and Mishra, 2004)

Figure 3: Indicators of Empowerment.

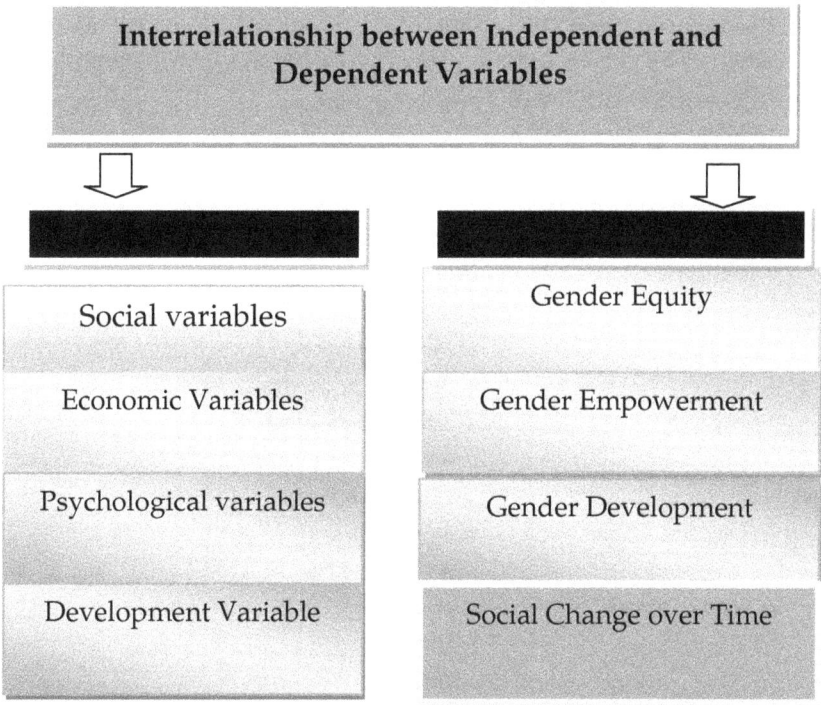

Figure 4: Interrelationship between Variables.

A. Independent Variables

I. Social Variables

Social life, society and social relationship have occupied the important position in social research. Social factors within which an individual resides and works are subject to change with time and space. A rural family looks intimately to the social factors in their livelihood management system. The empowerment of women living in rural areas is not independent of the following social variables. These are:

1. Marital status
2. Age of marriage
3. Choice of partner
4. Family size
5. Social participation
6. Social mobilization

Social Variables

1. **Marital status**: Operationally marital status refers to marriage of girls at specific age and freedom to choose spouse. In other words, the marital status may be considered as degree to which a girl exercises her freedom to choose life partner.
2. **Age of marriage**: This variable is to be measured as actual age at which the respondent got married. The prescribed age for girls to marry is 18 years.
3. **Choice of partner:** The variable can be measured in terms of mate selection, whether by personal choice, selection by parents or with negotiation.

Table 3: Scoring Pattern

Sl.No.	Major Item	Parameters	Score
1.	Marital status	Married	1
		Not married	0
2.	Age at marriage	Below 18 years	0
		Above 18 years	1
3.	Choice of partner	a. Personal choice	2
		b. Selection by parents	1
4.	Family size	a. 1-3 members (small)	1
		b. 4-7 members (medium)	2
		c. 7 and more (Large)	3
5.	Social participation	a. As member	1
		b. As office bearer	2
6.	Social mobilization	a. Agree	1
		b. Undecided	0
		c. Disagree	−1

Social participation = Score × Number of organization
Social mobilization = Number of statements × Score.

4. **Family size:** It is the concept that explains the total members of family who reside and share common kitchen. For quantification of response, the scores can be assigned as 1, 2, 3 and 4 for small family of up to three members, medium family within 4-6 members and large family beyond 7 members.

5. **Social participation:** It is the degree to which the individual women participate in social group, social function, social welfare programme and social institutions as member or office bearer.

6. **Social mobilization:** Statements representing social mobilization on a three point continuum of agree, undecided and disagree can be used to measure the empowerment of women in social mobilization.

II. Economic Variables

1. Farm size
2. Economic status
3. Income generating capacity
4. Family income
5. Economic productivity
6. Managerial ability

Economic Variables

1. Farm Size

It implies number of standard acres possessed by the respondents at the time of interview. The classification can be followed as:

Table 4: Farmers Category

Farmer Category	Irrigated Land (Acres)	Non-Irrigated Land (Acres)
1. Small farmer (up to 5 acres)		
2. Medium farmer (5-10 acres		
3. Big farmer (More than 10 acres)		

Scores assigned: 1, 2 and 3 for small, medium and big farmers 1 and 2 for Non-irrigated and Irrigated land X acres of land possessed.

2. Economic Status

It includes occupation of individual members of the family. Status is measured in terms of occupation and score can be assigned as:

Sl.No.	Occupation	Score
1.	Govt. service	5
2.	Agriculture	4
3.	Agriculture labour	3
4.	Subsidiary	2
5.	Other business	1

3. Income Generating Capacity

It refers to capacity of making economically and effective use of the available resources to maximize income level. The variable can be measured on a three point continuum like:

    a. Always

    b. Sometimes

    c. Never

We have to take some statements qualifying income generating capacity and score can be assigned as.

Sl.No.	Category	Score
1.	Always	3
2.	Sometimes	2
3.	Never	1

The total scores can be calculated over the statements and scores so obtained can be classified as low, medium and high managerial capability.

4. Family Income

The variables explain total earning by the entire family in a year from all sources, *i.e.,* farm and non-farm sources. One score can be assigned to each ₹1000 earning with categories of:

Sl.No.	Category	Score
1.	Low economic productivity	1
2.	Medium economic productivity	2
3.	High economic productivity	3

Sl.No.	Category	Score
1.	Less than ₹ 25,000	1
2.	₹ 25000 – ₹ 50,000	2
3.	₹ 50000 – ₹ 75,000	3
4.	₹ 75000 – ₹ 1,00,000	4
5.	More than ₹ 1,00,000	5

5. Economic Productivity

The variable explains as how women add to family income. The women are engaged in: (i) earning for family; (ii) household activities, and (iii) work outside home for payment. These three types of economic activities can be put in a continuum of: (a) most important, (b) contributes to some extent, (c) does not seem to be important at all and (d) not applicable with assigned score of 4, 3, 2 & 1.

Example

Sl.No.	Statements	Most Important	Contribution to some Extent	Does not Seem to Important at all	Not Applicable
1.	How is your contribution viewed in earning money for family				
2	How is your contribution as unpaid work you do				
3	How do you feel about yourself to work to earn money				
4.	How society views you to earn money for family				

In this case maximum and minimum obtainable score is 14 and 4. On the basis of scores the categorization may be made as:

Sl.No.	Category	Class Interval
1.	Low economic productivity	3-5
2.	Medium economic productivity	6-8
3.	High economic productivity	9 and above

6. Managerial Ability

The implied meaning of the variable "managerial ability "refers to performance of management functions. Within the livelihood sphere the individual performs the functions of management like, organizing, supervising, planning, communication and controlling the household activities. The variable can be studied under the headings of planning, organizing, supervising, communication, coordinating and controlling. The statements containing various aspects of managerial ability can be scored on three point continuum like, always, sometimes and never with assigned scores of 3, 2 and 1 respectively. On the basis of scores obtained the respondents can be classified as

Sl.No.	Category	Score
1.	Low managerial ability	1
2.	Medium managerial ability	2
3.	High managerial ability	3

III. Psychological Variables

These independent variables are:

1. Self orientation
2. Discrimination against girl child
3. Perception of drudgery

Psychological Variables

1. Self Orientation

Self orientation means degree of confidence of individuals in developing themselves. The variable can be measured qualifying statements on a three point continuum of agree, undecided and disagree with assigned score of 3,2,1 and for negative statements 1, 2 and 3 respectively. Adding of the scores over statements can lead to classification of sample respondents.

Sl.No.	Category	Score
1.	Low orientation	1
2.	Medium orientation	2
3.	High orientation	3

2. Discrimination against Girl Child

The variable explains perceptions of individuals against discrimination for girl child. The aim is to know whether women feel girl child is a burden to family. We can develop positive and negative statements to examine the perception for reliability. The staments can be rated on two point continuum of agree, and disagree with assigned score of 2 and 1 for positive and 1 and 2 for negative response. Thus the sample can be classified as:

Sl.No.	Category	Score
1.	Low discrimination	3
2.	Medium discrimination	2
3.	High discrimination	1

3. Perception about Drudgery

A normal individual works either in the farm or home for a period of 8 hours per day without feeling mental or physical strain (drudgery). The perception of drudgery can be defined as feeling of mental or physical strain for working 8 hours. The measurement of drudgery can be done on four point continuum of high, average, low and nil with assigned score of 4, 3, 2 & 1 respectively. On the basis of scores obtained the respondents can be classified as:

Sl.No.	Category	Score
1.	Low perception	1
2.	Medium perception	2
3.	High perception	3

IV. Development Variables

The developmental variables can be mentioned as:

1. Extension contact
2. Extension participation
3. Mass media exposure
4. Access to basic facility

Development Variables

1. Extension Contact

The variable refers to the degree to which respondent keeps contact with extension professional to obtain need based information. The extension professionals are Village Agriculture Workers, Agriculture Extension Officers and Agricultural Scientists. The contact can be measured in the terms of weekly, monthly, occasionally and never with assigned score of 4, 3, 2 and 1 respectively.

Sl.No.	Category	Score
1.	Low extension contact	1
2.	Medium extension contact	2
3.	High extension contact	3

2. Extension Participation

It is the frequency with which respondents are involved in extension activities like demonstration, group meeting, field day, farmer fairs etc. The variable can be measured in terms of regular, occasionally and never with assigned score of 3, 2 & 1. On the basis of obtained score they can be classified as:

Sl.No.	Category	Score
1.	Low participation	1
2.	Medium participation	2
3.	High participation	3

3. Mass Media Exposure

Mass media exposure is operationalized as the frequency of contact to various mass media in obtaining required information. The common mass media available are: news paper, radio, TV and farm magazines. The frequency of contact can be quantified as:

Sl.No.	Frequency	Score
1.	Almost daily	4
2.	Twice or thrice a week	3
3.	Once a week	2
4.	Never	1

On the basis of score obtained, the classification may be:

Sl.No.	Category	Score
1.	Low mass media exposure	1
2.	Medium mass media exposure	2
3.	High mass media exposure	3

4. Access to Basic Facility

Access to basic facilities like transport, housing, drinking water, environmental sanitation, power supply and recreation can be included in the variable. The score can be assigned 1 to each facility and classification may be as:

Sl.No.	Category	Score
1.	Low access	1
2.	Medium access	2
3.	High access	3

The independent variables are location specific and situation bound. The best way is to select variables as per objectives of the study adopting judge rating method specifying the variable as most relevant, relevant and not relevant with assigned core of 3,2 and 1 respectively.

B. Dependent Variables

Dependent variables in measuring empowerment are gender equity, gender empowerment and gender development. These three dependent variables together can provide a clear cut picture as to what extent the rural women have achieved empowerment.

1. Gender Equity

Gender equity refers to fair and equal deal with men and women. The review of literature reveals that gender equity means equal access of men and women to various resources that relevant to estimate equality. The identification of indicators can be done through judge rating methods. The judges are to be in field of women studies. The relevance rating normally recognizes five important indicators:

1. Education
2. Employment
3. Health and nutrition
4. Time use
5. Access and control over resources

1. Education

Education has the maximum effect on empowerment and constant efforts are made to provide education to women. Access to education includes enrolment in school, type of school, distance of school from the residence, transport facilities and opinion of family members for education of girls. These may not be exclusive and factors like educational climate of the schools can be included for more information. The opinion of respondents are to be properly rated on women education. For the purpose positive and negative statements can be framed against which the responses are to be recorded. For positive statements, the scores can be assigned as 3, 2 and 1 for agree, undecided and disagree and the procedure be reverse in case of negative statements. The scores obtained on each item be added to get overall opinion of the respondent about women education.

2. Employment

Access to employment is another indicator of empowerment. It includes employment opportunities in the village, migration, type of employment, number of man days generated within and outside village, mobility, awareness about various employment schemes setting up of enterprise. Number of man days generated can be quantified by counting actually how many days are available for employment. The simple questions on employment can be prepared with dichotomous pattern agree and disagree or yes and no to get overall score on accessibility to employment. Positive and negative statements qualifying access to employment can be rated on agree, undecided and disagree with score of 3, 2 and 1 and reverse for negative statements. Women are given work on gender basis. To measure employment aspects the scores can be assigned as follows:

Sl.No.	Category	Score
1.	Men only	1
2.	Men and sometimes women	2
3.	Both	3
4.	Men women and sometimes men	4
5.	Women only	5

3. Health and Nutrition

The dependent variable includes accessibility of women to health care and nutrition in terms of health facility, availing of health facilities, food consumption, nutrient intake and nutritional status. Three components of the dependent variables are: (i) Access to health (ii) Availing of health care and (iii) Food consumption pattern.

(i) Access to Health

1. Doctor and medical facility
2. Use of available health services
3. Starting of treatment as and when ill
4. Aware of health programs
5. Attending of health campaign
6. Problem faced in hospital
7. Pre and post natal care
8. Food and accommodation in hospital

(i) Availing Health Services

1. Equal medical facility for men and women
2. Better medical care and treatment
3. Women are more susceptible to illness for which more care given
4. Treatment before severe condition
5. More medical care needed for women

Sl.No.	Frequency	Score
1.	Daily	4
2.	Weekly	3
3.	Monthly	2
4.	Not at all	1

(iii) Food Intake items	Do not take at all	Daily	Weekly	Monthly
1. Cereals				
2. Pulses				
3. Milk				
4. Egg				
5. Fish				
6. Meat				
7. Fruits				

Further attempt can be made to find out nutrition intake in terms of: (i) Calorie intake, (ii) Protein intake, (iii) Nutritional status taking Height, Weight and BMI.

4. Time Use

The variable covers the areas like man days devoted in respect of farming, livestock rearing and household activities. The normal activities of rural women in respect of farming, livestock rearing and household duties are furnished below, but these are location specific which provides scope for addition or deletion of items.

Farm Related Items	**Livestock Rearing**	**Household Activities**
1. Land preparation	1. Bringing fodder	1. Food preparation
2 Nursery raising	2. Preparing feed	2. Feeding children
3. Transplanting	3. Feeding	3. House cleaning
4. Manure appli-cation	4 Grazing	4. Marketing
5 Intercultural activities	5. Cleaning/bathing animals	5. Treatment at illness
6. Irrigation	6. Health care	6. Guest entertain-ment
7. Plant protection	7. Cleaning sheds	7. Social function
8. Harvesting	8. Watering	8. Cleaning of cloths
9. Post harvest care	9. Milking	9. Bringing water
10. Marketing	10. All livestock activities	10. Care of children

Time use of the rural women basing on the response on the items indicated above, SD and mean can be calculated and categorization may be made as: (i) Low, (ii) Medium and (iii) High time users.

5. Access and Control over Resources

The concept implies that how the resources like, land, labour, credit and income are allocated between men and women. Control is conceptualized as allowing a person to take decision about who uses resources or disposes resources.

Access and control over resources can be studied covering a wide range of items. The list is given below.

Table 5: Access and Control Over Resources

Sl.No.	Resources	Access		Control	
		Men	Women	Men	Women
1.	Land				
2.	Credit				
3.	House				
4.	Immovable properties				
5.	Consumption of household items				
6.	Tools and machineries				
7.	Livestock				

Contd...

Table 5–*Contd...*

Sl.No.	Resources	Access		Control	
		Men	*Women*	*Men*	*Women*
8.	Income				
10.	Employment				
11	Education				
12.	Production input				
13.	Production output				

Assigning score of 2 and 2 for access and control over resources against 1 and 1 of no access and no control, we can classify the respondents into low, medium and high. In these cases maximum scores will be 26 + 26 = 52 and minimum 13 + 13 = 26.

Sl.No.	Category	Access	Control
1.	Low	Up to 17	Up to 17
2.	Medium	18-20	18-20
3.	High	21 and above	21 and above

Gender Equity Index

To measure gender equity, weighted index can be developed. To work out Gender Equity Index we have to use Judge Rating method. We have to decide number of judges and areas of their specialization or concern or working in. Each judge will be requested to rate the relevancy of the indicator for equity and score them out of 100. Taking total scores obtained for each item rated by judges and dividing by number of judges, we can obtain weightage.

Example

Minimum number of judges should be preferably 25-30. Suppose we have taken 5 judges and they have rated items like as given in Table 6.

Table 6: Gender Equity Index

Sl.No.	Indicators of Equity	J1	J2	J3	J4	J5	Total/5	Weightage (Total/5/10)
1.	Education	20	5	15	20	12	14.4	1.4
2.	Employment	10	5	10	15	18	11.6	1.2
3.	Health and nutrition	30	60	35	30	30	37	3.7
4.	Time	20	20	20	25	20	21	2.1
5.	Access and control over resources	20	10	20	10	20	16	1.6
	Total	**100**	**100**	**100**	**100**	**100**	**100**	**10**

The weightage score can be used for correction and any other statistical treatment.

Gender Empowerment Index

Out of the past studies and situation for which the study is designed can use five indicators to develop gender empowerment index.

These are:

1. Decision-making
2. Social empowerment
3. Economic empowerment
4. Political empowerment
5. Legal empowerment

Operational definition and measurement of empowerment indicators.

1. Decision-making

Decision-making is a mental process based on conscious reasoning (Reick, 1960). Roger (1967) further clarified that decision-making is a process by which an evaluation of the meaning and consequence of alternative one of conduct is made. In rural areas the decision-making areas mostly as follows.

Table 7: Areas of Decision-making

Household Activities	Farm activities	Financial decisions	Socio-religious decision
1. Home management	1. Agriculture	1. Money management	1. Household exchange
2. Child related decisions	2. Labour allocation	2. Capital transaction	2. Short distance travel
	3. Livestock related	3. Purchase of farm inputs	3. Socio-religious obligation
		4. Disposal of farm produce	
		5. Disposal of home produce	

The decision-making process is measured in (where women takes full decision = 3, with men as joint = 2 and only men = 1. Under each heading, household, farm, financial and socio-religious decision a number of items can be included.

Home Related Decisions

(i) Cooking particular food
(ii) Construction of new house
(ii) Repair of existing house
(iv) Interior decoration
(v) Miscellaneous home management decision

Child Related Decision

   (i) Purchase for children

  (ii) Treatment of children

 (iii) Age of schooling

 (iv) Level of education

 (iv) Type of education

  (v) Marriage of children

 (vi) Selection of match

(vii) Marriage expenses

Farm Related Decision

   (i) Kind of crop

  (ii) Area under each crop

 (iii) Seed and variety selection

 (iv) Fertilizer use

 (iv) Irrigation

  (v) Plant protection

 (vi) Harvest and post harvest operation

Labour Allocation

   (i) Arranging labour

  (ii) Deciding work for labour

 (iii) Wage of labour

Livestock Related Decision

   (i) Type of livestock to keep

  (ii) Number of livestock

 (iii) Feeding

 (iv) Vaccination

 (iv) A.I.

  (v) Treatment of animal

Financial Decision

   (i) Money management, keeping money and how much to spend

  (ii) Buying/selling land

 (iii) Buying/selling livestock

 (iv) Purchase of farm input, implement

 (iv) Buying of food material, cloth, household durable

 (vi) Disposal of farm produce, selling of food grains, animals

  (v) Attending marriage function

    (vi) Giving donation

   (vii) Religious functions

  (viii) Decision on loan and repayment

    (ix) Saving/bank

    (x) Buying/selling jewelry

Household Exchange

    (i) Small loan

   (ii) Grain/vegetables

  (iii) Gift

Short Social Travel

    (i) Mela/Exhibition

   (ii) Courtesy call

Socio-Religious Obligation

    (i) Daughter and in laws

   (ii) Attending marriage/death ceremonies

  (iii) Charity

  (iv) Donation to school

  (iv) Religious meeting

Calculation of scores over all these items can be made to arrive possible range and categorization be made accordingly.

II. Social Empowerment

Social empowerment means autonomy of women in respect of physical mobility, dominance, control by men, mental tension and difference with partner.

The measurement of social empowerment may be done following three point rating continuum like always, sometimes and never with assigned value of 3, 2 and 1. The areas of social empowerment are:

1. Taking of independent initiatives
2. Mobility
3. Contact with own parents and relatives
4. Freedom to visit parental home
5. To do social work
6. Threat of violence
7. Freedom to take independent decision
8. Freedom to attend social functions with others
9. Getting prestige at home
10. Participation in family discussion

11. Opinion on expenditure
12. Freedom for child bearing decision
13. Can vote without influence of family
14. Free from fear of marital break up

III. Economic Empowerment

Economic empowerment refers to autonomy on the matter of spending money, investment, personal income, participation in household expenditure, custody of ornament and saving. The scoring pattern can be adopted as 3, 2 and 1 for always, sometimes and never respectively.

1. Non-dependence on husband to meet economic needs
2. Allowed to spend as per need
3. Consultation before investment
4. Consultation on expenditure
5. Freedom on personal income
6. Independent saving
7. Allowed to purchase jewelry
8. Permitted to sell/trade land
9. Freedom to make purchase/negotiation
10. Takes part in decision for heavy expenditure
11. Consulted on loan and repayment
12. Knows well about financial deals of family

The response against items of economic empowerment can be quantified and range of scores be calculated to indicate level of empowerment. The aim is to determine whether women are economically empowered or at par with counterpart. The maximum and minimum score range can provide basis for categorization of respondents into high, medium and low economic empowerment level.

IV. Political Empowerment

Political empowerment is a complex subject. But to quantify the concept, there is need to find out indicators. The indicators are: (i) political awareness, (ii) extent of political participation and opinion about women involvement in politics.

(a) *Political awareness*: It can be ascertained in terms of political situation at national, state, district and block level, election process, names of political leaders and political parties.

(b) *Political participation*: The political participation may be measured in terms of free to caste vote, tasking up of political campaign, support of political parties etc. The continuum can be used in three points like, always, sometimes and never with assigned scores of 3, 2 and 1.

(c) *Opinion about women in politics*: The feeling of the sample may be measured on a attitude scale of three points like, agree, undecided and disagree with a scoring pattern of 3, 2 and 1.

(d) The total scores on account of three areas of awareness, participation and opinion can provide total picture as where do women stand in political empowerment.

The items of measurement are:

1. Awareness Indicators

1. Understanding of national and state government
2. Nature of government
3. Names of political parties
4. Election for MP, MLA
5. Names of popular leaders
6. Political problems of the nation and state
7. Local political problems
8. Process of election
9. Voting process
10. Political campaigns

   (response may be secured on Yes or No)

a. Political Participation

1. Level of interest in politics
2. How often discuss about political matter
3. Interest in political campaign
4. Openly support to political leaders
5. Participation in voting
6. Attending of political meetings
7. Participation in political decision-making process at local level
8. Interest in contesting election at any level
9. Taking part in distribution of political literature
10. Any other political activity

    (For each item response may be recorded differently in two or three point continuum)

b. Opinion about Women in Politics

1. Good domain for women
2. Vote as per decision of male members
3. Women to contest for betterment of society

4. Political participation gives status and prestige
5. Participation in politics means fight against social evils
6. Consultation with women for politics is improvement of situation

(Statements may be formed on these items and responses be recorded in terms of agree, undecided and disagree)

V. Legal Empowerment

Legal empowerment means awareness and opinion about legal acts for safeguard of women in society. Awareness may be scored as 2 and 1 for Yes and No and opinion about legal acts in safeguarding women in a three point continuum as agree, neutral and disagree:

1. Awareness about legal acts
2. Child restraint act
3. Divorce act
4. Dowry prohibition act
5. Widow marriage act
6. Right to property act
1. Right to information act
7. Rape related act
8. Family violence act
9. Right to services act
10. Opinion about legal act

Statements relating to the acts cited above may be constructed and responses be secured on agree, undecided and disagree with scoring of 3, 2 and 1 for positive and reverse for negative response. All the scores obtained on these aspects can be used for categorization of respondents into high, medium and low opinion groups.

VI. Gender Development Index

As per United Nations Development norm, the gender development index is to include:

1. Life expectancy
2. Education
3. Income

The Gender Development Index includes the dimensions as stated below.

1. Long and healthy life as measured by life expectancy at birth.
2. Education means adult literacy rate combined of primary, secondary and tertiary gross enrolment ratio.
3. Income means decent standard living measured by estimated earned income.

Dependent variables in totality are: Gender Equity Index, Gender Empowerment Index and Gender Development Index

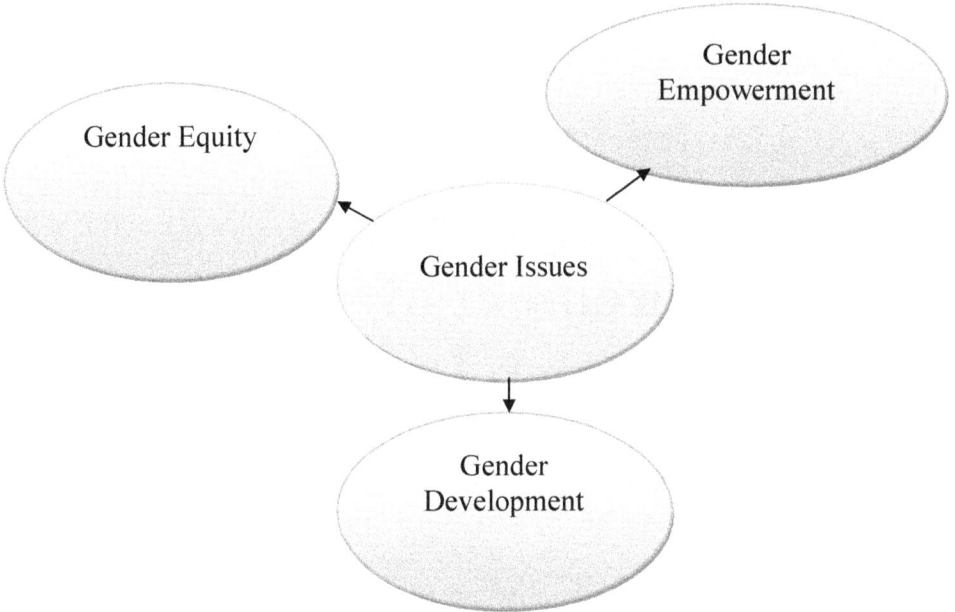

Figure 5: Dependent Variables for Gender Equity.

Chapter 3

Measurement of Traits of Women Leaders

"Can women lead" was the question among the scholars and social scientists for a long time. Now of course it is past. There were some view points about the efficiency of women taking leadership position. The differences in view points made to believe that women are inferior to men to perform leadership role. The increasing number of women in leadership position at present brought about dramatic changes in our society. Now the question is whether women lead in different manner from men and whether women or men are more effective. Eagly and Johnson (1990) in their meta analysis found contrary to stereotypic expectation that women were not found to lead in a more interpersonally and less task oriented manner than men. The difference was attributed to social system where social rules are more regulated. Another Meta analysis by Van Engen (2001) revealed that women lead in a more democratic or participative manner than men.

The corporate houses and management organizations have proved through research that prejudice about leadership role of women has no basis. The effectiveness of women leadership is explained in cycle as shown in Figure 6.

From qualitative synthesis of earlier research Kirkpatrik and Locke (1991) postulated that leadership differs from non-leaders in six important traits. These are:

1. Drive
2. Desire to lead
3. Honesty and integrity
4. Self confidence

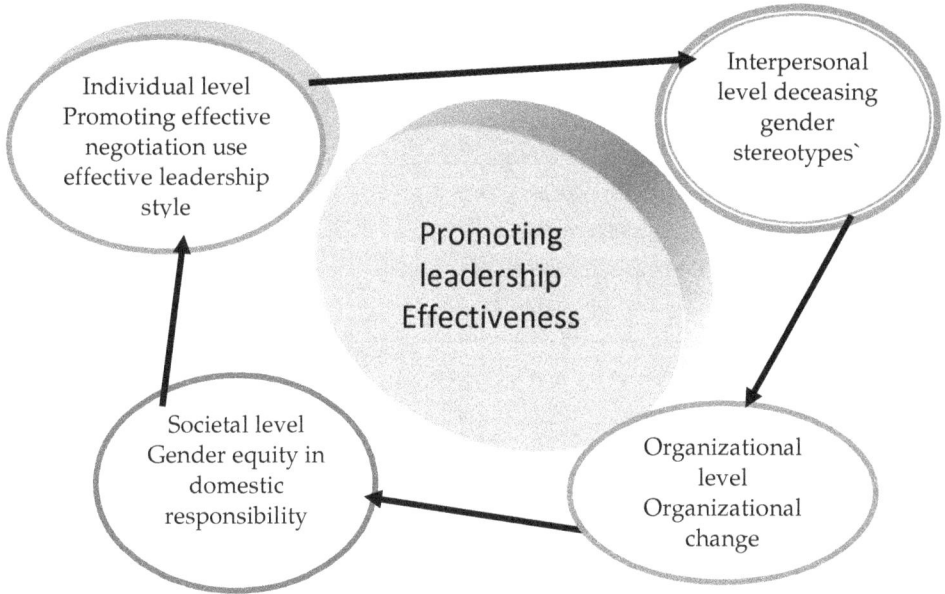

Figure 6: Promoting Leadership Effectiveness.

5. Cognitive ability
6. Knowledge of business

According to some school of thoughts individuals can be born without these traits and can learn to make difference. There are about four methods to study leadership and more specifically women leaders.

Four Approaches of Leadership Study
1. **Trait Approach**
2. **Skill Approach**
3. **Style Approach**
4. **Situational approach**

I. Trait Approach

A number of authors have worked on leadership and prominent among them are: Stogdill (1948), Mann (1959), Stogdill (1974), Lord, DeVdar and Alliger (1986) and Kirkpatric and Locke (1991). These authors have found out different traits of leadership pattern. These are presented in box showing in Table 8 and Figure 7.

Looking at the traits postulated by different authors, altogether six traits have been accepted at international level without any controversy. These are:

1. Intelligence
2. Self confidence
3. Determination
4. Integrity
5. Sociability

1. Strongly agree	5
2. Agree	4
3. Neutral	3
4. Disagree	2
5. Strongly disagree	1

These indicators can be measured on a five point continuum and the score assignment may be as follows.

Table 8: Traits of all Indicators

Sl.No.	Traits	1	2	3	4	5
1	Articulation: Communicates effectively with others					
2	Perception: Discerning and insightful					
3	Self confidence: Believes in oneself and one's ability					
4.	Self assured: Secure with self, free of doubts					
5	Persistent: Stays fixed on goals despite interference					
6	Determined: Takes firm stand, acts with certainty					
7	Trustworthy: Acts believably, inspires confidence					
8	Dependable: Is consistent and reliable					
9	Friendly: Shows kindness and warmth					
10	Outgoing: Talks freely,gets well along with others					

Source: Leadership Theory and Practice, Northouse, Sage Publication, 2007, p. 33.

Against each item the response can be obtained and scores be assigned to find out relative position of each trait and accordingly conclusion can be drawn about the leader to put the instrument into action and calculation of score may be as: (Study in Odisha 2009).

Table 9: Relative Position of Traits of Leadership

Traits	R1	R2	R3	R4	R5	Average	Self Assessment	Difference
1. Articulate	4	3	5	2	5	3.8	5	-1.2
2. Perceptive	3	5	5	5	4	4.4	3	+1.4
3. Self confidence	4	4	5	4	4	4.2	4	+0.2
3. Self assured	3	2	4	4	5	3.60	4	-0.40
4. Persistent	5	4	1	2	3	3.00	2	-+1.00
5. Determined	4	5	4	2	3	3.60	3	+0.60
6. Trust worthy	2	5	3	3	4	3.40	4	-0.60
7. Dependable	5	4	3	2	5	4.00	3	-1.00
8. Friendly	4	4	2	5	3	3.60	3	+.0.60
9. Outgoing	5	3	3	4	4	3.80	3	+.8-0

The score for each trait can be calculated, self assessment of the respondents can be secured and analysis be made to draw conclusion. Total outcome *i.e.,* + and – will indicate.

II. Skill Approach

The approach is leader-centered perspective. The skill approaches are of three types:

1. **Technical Skill:** Technical skill is the proficiency in a specific type of work. It includes competencies in a specialized area, analytical ability and ability to use tools and techniques.
2. **Human Skill:** It is knowledge about and ability to work with people. Human skills allow leader to assist group members in working cooperatively as a group to achieve common goal
3. **Conceptual Skill:** It is the skill to work with ideas and concepts. A leader with conceptual skill is comfortable talking about idea that shapes the organization.

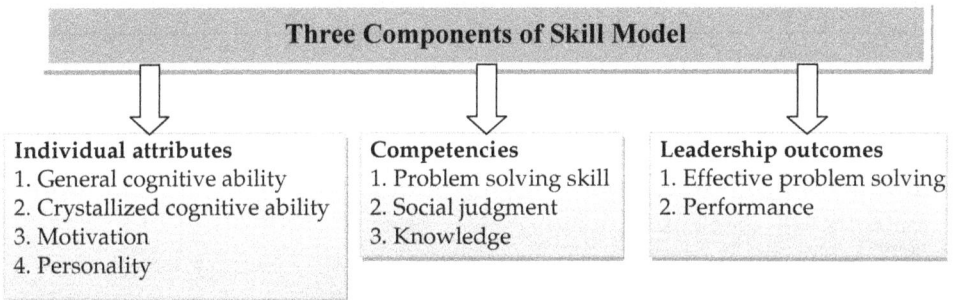

Three Components of Skill Model		
Individual attributes 1. General cognitive ability 2. Crystallized cognitive ability 3. Motivation 4. Personality	**Competencies** 1. Problem solving skill 2. Social judgment 3. Knowledge	**Leadership outcomes** 1. Effective problem solving 2. Performance

Figure 8: Components of Skill Model.

The skill approach covering technical, human and conceptual can be measured assigning score against the statements qualifying these aspects.

1. Not true	1
2. Seldom true	2
3. Occasionally true	3
4. Somewhat true	4
5. Very true	5

Table 10: Technical Skill

Sl.No.	Technical Skill	5	4	3	2	1	AV
1.	Enjoy to know how the work is done	6	5	4	3	7	3.00
2.	Technical things bring fascination	3	8	7	4	3	3.16
3.	One of the skill is to make things work	10	5	6	3	1	3.80
4.	Following direction to do the work is easy	3	8	10	4	0	3.40
5.	Competing with others as per assignment	2	4	5	13	1	2.88

A list of statements describing Technical, Human and Conceptual are given herewith along with pattern of scoring. A study conducted in Odisha to know the leadership pattern among the members of SHG with a sample of 25, in 2007 covering blocks of Puri district.

Table 11: Human Skill

Sl.No.	Human Skill	5	4	3	2	1	AV
1.	Adopting idea according to needs of the people	4	3	8	5	5	2.72
2.	Being able to understand others is most important part of work	3	2	10	3	7	2.64
3.	Supportive communication is a must	4	6	8	5	2	2.96
4.	Getting all to work together is a challenge	5	2	8	5	2	2.76
5.	Interest to know how the decision affects life of people	3	6	6	5	5	2.88

Table 12: Conceptual Skill

Sl.No.	Conceptual Skill	5	4	3	2	1	AV
1.	Like to work with abstract ideas	3	6	3	8	5	2.76
2.	Big pictures come to us	4	7	8	3	3	3.24
3.	Intrigued by complex organizational problems	3	7	10	5	0	3.32
4.	Enjoy work with strategy of work	6	8	7	2	2	3.64
5.	Mission based work is rewarding	9	6	5	4	1	3.72
6.	Organizational value and philosophy is important	7	12	2	3	1	3.84

Indicators of Leadership traits (Northouse, Theory and Practice, Sage Publication, 2007, p. 33).

Total scores obtained can be indicated for technical, human and conceptual skill. The score provides information about the strength and weakness of the concerned leaders and accordingly they can fill up their deficiencies.

III. Style Approach

The style approach emphasizes the behaviour of the leader. It focuses on what leaders do and how they act. There are five leadership style approaches.

1. **Authority-Compliance**: Much more emphasis on task and job requirements and less emphasis on people. It is result driven approach where communication is only instruction.

2. **Country club management**: The approach is more concerned with interpersonal relationship and less concern or low concern with job. Leaders look for attitude and feeling of people making sure that personal and social needs of people are met.

3. **Impoverished management**: Leaders are unconcerned with job and interpersonal relationship. Leaders do not get involved.

4. **Middle of the road management**: This type of approach emphasizes intermediate concern. They try to bring a balance taking people into account and emphasizing job requirements. The leaders try to avoid conflicts.

5. **Team Management**: It promotes high degree of participation and satisfy basic needs of the people emphasis on job and interpersonal relationship.

Table 13: Style Approach

Sl.No.	Style Approach	Job Requirements	Interpersonal Relationship
1.	Authority compliance	Much emphasis	Less emphasis
2.	Country club management	Less	Much more
3.	Impoverished management	Less	Less
4.	Middle of the road management	Equal emphasis but not strong	Equal emphasis but not strong
5.	Team Management	Strong emphasis	Strong emphasis

1. Never	1
2. Seldom	2
3. Occasionally	3
4. Often	4
5. Always	5

Measurement

The style approach of leadership can be measured on a five point continuum with scores as follows.

The statements qualifying style approach can be framed and the respondents can respond and the scores can be calculated.

Statements may be on:

Task Criteria

1. Tell group members what supposed to do
2. Set standards of performance
3. Help others to feel comfortable
4. Respond favourably to suggestions of others
5. Make perspective clear
6. Develop plan of action
7. Clarifies role
8. Provide plan to complete work
9. Provide criteria for what is expected
10. Encourage for high quality work

Relationship Criteria

1. Act friendly
2. Help others to feel comfortable
3. Respond favourably to suggestions of others
4. Treat others fairly
5. Behave in predictable manner
6. Make communication about activities
7. Shows concern of well being
8. Flexibility in making decision
9. Disclose own thoughts and feelings
10. Helps group members get along

Classification of leaders on style approach

Range of Score	Level
45-50	Very high
40-44	High
35-39	Moderately high
30-34	Moderately low
25-29	Low
10-24	Very low

IV. Situational Approach

One of the most widely recognized approaches to leadership is the situational approach. It focuses leadership in situations. Situational approach emphasizes the behaviour of the leaders. Situational leadership consists of behavioural pattern of a person who attempts to influence others. It includes (i) Directive (task) and supportive (relationship) behaviour. The task behaviour facilitates goal accomplishment and help group members to achieve their objectives. Relationship behaviour helps subordinates to feel comfortable with themselves with each other and situations in which they work.

1. Directive or Task Approach Emphasizes

a. Achieve goal by giving direction
b. Establishing goal to achieve
c. Decide method of evaluation
d. Setting time limit to complete work
e. Defining role of each member
f. Giving demonstration as how to achieve goal
g. One way communication is emphasized

2. Supportive or Relationship Style of Leadership Emphasizes

    a. Feel comfortable with people and situation

    b. Relation with co-workers

    c. Two way communication

    d. Emotional support to members

Further situational leadership style can be grouped into four distinct categories.

Table 14: Situational Leadership Style

Sl.No.	Category	Directive	Supportive
1.	Directive style	High directive	Low supportive
2.	Coaching approach	High directive	High supportive
3.	Supporting approach	Low directive	High supportive
4.	Delegating approach	Low directive	Low supportive

Table 15: Imagining Leaders by Men and Women

Women Imagining Women as Leaders	Men Imagining Women as Leaders	Men Imaging Men as Leaders
1. Mobility	1. Strength to work	1. Can write
2. Need to work hard	2. Motivate people	2. Ability to collect work
3. Be active	3. Good character	3. Ability to mobilize
4. Good character	4. Must smell good	4. Good behaviour
5. Good vision	5. Follow order of husband	5. Good health
6. Treat everyone equally	6. Intelligence	6. Discuss with people
7. Intelligence	7. Educated	7. Regular meeting
8. Education	8. Calm and composed	8. Motivate people
9. Patience	9. Numerical skill	9. Thinking power
10. Cooperation	10. See all are equal	10. Educated
11. Humbleness	11. Communication skill	11. Good listener
12. Good listener	12. Patience	12. Communicative
13. Bold to talk	13. Courage	13. Able to guide
14. Courage	14. Keep house clean	14. Decision-making power
15. Unify members	15. Carry launch for husband	15. Bold and determined
		16. Formulate agenda
		17. Strength to work
		18. Mobility

Measurement of situational approach is little different than others. For each work performed or to be performed, the scores can be assigned in four dimensions. The four possible responses can be obtained in cell given above. Which style of approach can be determined such as: directive, coaching, supportive or delegative and conclusions be drawn accordingly.

Place of Women in the Domain of Leadership

Baumgartner and Hogger (2004), their 'in search of livelihood' have tried to study women in the context of leadership in five contexts like, individual women and families, street level SHG, village WIC level, panchayat level, gram panchayat and district level(DWMC).They used visual imaging exercise with group of women and men at different forum. The exercise that women imaging women as leaders yielded the some desirable qualities or traits of women leaders. On other hand, men were included in separate exercise to image women as leaders also brought out some desirable qualities; while men tried to image men as leaders, a great difference was observed.

Comparative Imaging of Women and Men about Leadership Traits

Leadership qualities stated by both women and men are:

1. Active
2. Treat followers well
3. Educated
4. Bold
5. Ability to move around
6. Determined
7. Patient
8. Good character/behaviour

The qualities of women leaders can be measured in a three point continuum like, very much, much and little assigning the scores of 3, 2 and 1. Summation of scores divided by sample frequency can provide the average mean to rank the women leaders.

Further, World Bank in Gender Equality and Development, 2012, p. 85 mentioned the world scenario of perceived leadership. "Men are perceived as better political leaders than women" illustrates the worldwide figures obtained of our survey as follows.

The leadership traits of men and women are differently perceived at different countries. The World Bank survey presented interesting facts about adolescents. Across rural communities in eight countries, 800 boys and girls between age group of 11 and 17 in focus group discussions expressed the traits of good and bad boys and girls.

The qualities of women leaders can be measured in a three point continuum like, very much, much and little assigning the scores of 3, 2 and 1. Summation of scores divided by number of respondents can provide the average mean to rank the women leaders.

Table 16: World Scenario of Perceived Leadership

Sl.No.	Countries	Per cent of People who Believe that Men Make Better Political Leaders than Women do	Sl.No.	Countries	Per cent of People who Believe that Men Make Better Political Leaders than Women do
1.	Georgia	68.00	2.	India	63.00
3.	Russia Federation	62.00	4.	Turkey	66.00
5.	Romaina	67.00	6.	Ukraine	63.00
7.	China	54.00	8.	South Africa	51.00
9.	Chile	42.00	10.	Bulgaria	48.00
11.	Japan	44.00	12.	Poland	43.00
13.	Brazil	32.00	14.	Slovenia	31.00
15.	Mexico	28.00	16.	Uruguay	20.00
17.	Peru	18.00	18.	Sweden	8.00

The World Bank, 2012, Gender Equality and Development, Washington DC 20433, p. 85.

Good Girl

1. Polite
2. Hard working
3. No boy friend
4. Stays home
5. Dress appropriate
6. Religious
7. Obedient
8. Respectful
9. Helps at home
10. Studies

Bad Girls

1. Fights
2. Prostitute
3. Boy friends
4. Does not listen to parents
5. Vices
6. Disobedient
7. Not religious
8. Does not help at home
9. Disrespectful
10. Dress inappropriate
11. Does not study
12. Roams around

Good Boy

1. No girl friend
2. Has good friends
3. Polite
4. Obedient
5. Religious
6. Helps at home
7. No vices
8. Respectful
9. Studies

Bad Boy

1. Lies
2. Fights
3. Many girl friends
4. Disobedient
5. Not religious
6. Steals
7. Disrespectful
8. Roams around
9. Does not study
10. Vices

In the present context, the leadership pattern in mushrooming SHGs needs study and implementation of ideals. The SHGs are village level units which can deliver goods to remove poverty and inequality. If correct and desirable directions are given to the SHGs they will serve as main channels of development.

Leaders always function in group. The group and their leaders operate under influence of multifactor.

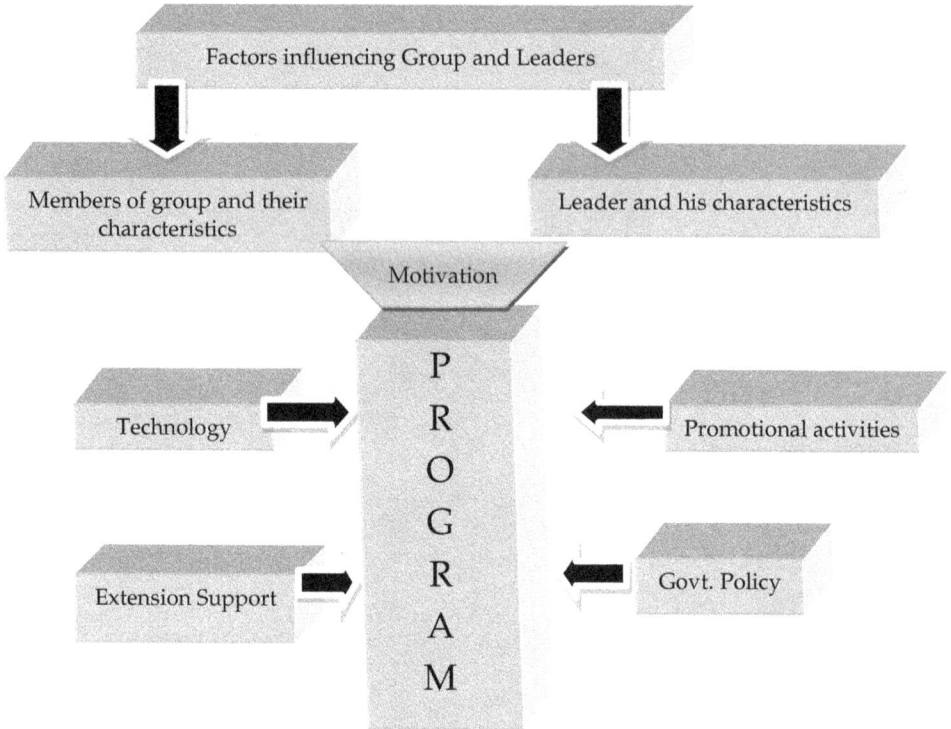

Figure 9: Factors Influencing Groups and Leaders.

Status of SHG

Table 17: Region-wise Cumulative Number of SHGs in India

Sl.No.	Region/State	No. of SHGs	Per cent of SHGs
1.	Northern regions	170,059	5.05
2.	Northern-Eastern region	132,111	3.92
3.	Eastern region	631,958	18.76
4.	Central region	722.846	21.46
5.	Western region	298,598	8.876
6.	Southern region	1,412,643	41.94

Source: NABARD progress report of SHG- Bank linkage in India-2006.

Table 18: Region-wise SHG Federation in India (March 2007)

Sl.No.	Region	No. of SHG Federation
1.	Northern Region	147
2.	North-Eastern Region	132
3.	Eastern Region	5,825
4.	Central Region	821
5.	Western Region	664
6.	Southern Region	61,286
	All India	**68,903**

Table 19: Self Help Group in Odisha (April, 2010)

Sl.No.	Areas of Highlight	No/₹
1	No. of women SHG	4,15,203
2.	Membership	49,82,439
3.	Credit advanced (lakhs)	₹ 1,59,582,70
4	No. SHG credit linked	4,80,045
5.	No. of Federations formed	7838

Table 20: District-wise Women Self Help Groups (2005)

Sl.No.	District	No. of Groups	Per cent of Groups	No. of Members	Per cent of Membership
1	Angul	6508	4.27	76949	4.00
2.	Balasore	7344	4.82	90177	4.68
3.	Baragarh	7374	4.84	76496	3.97
4.	Bhadrakh	4670	3.06	62044	3.22
5.	Bolangir	10159	6.67	123170	6.39
6.	Boudh	1986	1.30	25909	1.34
7.	Cuttack	6399	4.29	84137	4.37
8.	Deogarh	1282	0.84	17724	0.92
9.	Dhenknal	6680	4.43	81230	4.22
10.	Gajapati	2256	1.48	30660	1.59
11.	Ganjam	13646	8.95	171548	8.92
12.	Jagatsinghpur	5432	3.56	73531	3.84
13.	Jajpur	4397	2.88	63152	3.28
14.	Jharsuguda	1761	1.16	20970	1.09
15.	Kalahandi	4604	3.02	53534	2.78
16.	Kendrapara	3980	2.61	57837	3.00

Contd...

Table 20–*Contd...*

Sl.No.	District	No. of Groups	Per cent of Groups	No. of Members	Per cent of Membership
17.	Keonjhar	4888	3.20	54122	2.83
18.	Khurda	2995	1.96	38097	1.98
19.	Koraput	4246	2.79	54603	2.84
20.	Malkangiri	6759	4.43	80302	4.17
21.	Mayurbhanja	12427	8.15	142970	7.43
22.	Nawarangapur	2094	1.31	26476	1.37
23.	Nayagarh	1848	1.20	22162	1.15
24.	Nuapada	3741	2.44	47081	2.44
25.	Phulbani	2335	1.52	27431	1.42
26.	Puri	7046	4.61	108846	5.65
27.	Raygada	5650	3.70	73191	3.80
28.	Sambalpur	3468	2.26	42795	2.22
29.	Sonepur	2035	1.32	23241	1.21
30.	Sundargarh	6369	4.17	74681	3.88
	Total	**152379**	**100.00**	**1925075**	**100.00**

Chapter 4

Measurement of Capacity Building and Knowledge Gain

The efforts to empower rural women have been the focus of state and national governments. It is an established fact that empowerment needs knowledge, skill and positive attitude. The training and capacity building are used to induce knowledge and skill in human being to be positive to development projects. Training of Rural Women has resulted in good adoption trend everywhere. Sometimes we neglect to assess as how much knowledge has been gained by the participants to formulate future plan of action.

There are two important concepts that we use are capacity building and training. Both the concepts are used to induce knowledge, skill and improved skill with the participants.

1. Capacity Building

The interpretation of capacity building focuses in promotion of development mostly in developing countries. Training center, learning center and consultants are all some form of capacity building. The UNDP used the concept capacity building in the context of development. Since early 1970s the UNDP offered guidance for its staff and governments on what was considered "institution building". In 1991 the term was used for community capacity building. The UNDP defines capacity building as "long term continual process of development that involves all stakeholders, including ministries, local authorities, NGOs, professionals, community members, academies and more." Capacity building uses scientific, technological, organizations and institutional resource capabilities. The aim of capacity building is to tackle the problems.

Three Levels of Capacity Building

Capacity building

⇓   ⇓   ⇓

Individual level	**Institutional level**	**Societal Level**
1. Enhance knowledge	1. Modernizing existing institution	1. Interactive public administration
2. Skill	2. Support to frame sound policy	2. Feedback analysis
3. Process of learning	3. Organizational structure	3. Responsive
4. Adapting to change	4. Effective management system	4. Accountable
	5. Revenue control	

Figure 10: Capacity Building.

The World Customs Organization which is an intergovernmental organization views capacity building as activities which strengthen the knowledge, abilities, skill and behaviour of individuals and improve institutional structure and processes such that the organization can efficiently meet its mission and goal in a sustainable way. OXFAM an International a globally recognized NGO looks community capacity building as an approach to development. Allan Kaplan a leading NGO argues to be effective facilitator of capacity building is the acquiring of skills and resources for developing a vision strategy and organizational structure.

1. Engage stakeholders on capacity development
2. Assess capacity needs and assets
3. Formulate a capacity development response through:
   a. Institutional arrangement
   b. Leadership
   c. Knowledge
   d. Accountability
4. Implement a capacity development response
5. Evaluate capacity development

In the UNDP 2008-2013 strategic plan for development based on the concept of capacity building suggests the steps like, synthesis of above definitions and concepts lead to understand that human development can be achieved through training in which knowledge, skill and attitude have to be given top priority.

2. Training

There are numerous definitions of training. Different authors have defined training according to their situation and context. But the appropriate definition of training is that, "training is a planned communication process which results in changes in attitude, knowledge and skill in accordance with specified objectives relating to desired patterns of behaviour (M. Khemmani, 1983).

In the fields of agriculture, animal husbandry, fisheries and forestry, trainings are offered by many agencies suiting to their requirements. The result is assessed in terms of adoption of practices that are included in training program. Experiential learning cycle is mostly used in training programme to have more retention of knowledge by the participants. Among baskets of training methods, experience of the participant is given more emphasis to yield better results. Experiential training model recognizes identification of goals in the line of experiential learning cycle. The steps in experiential learning cycles are,

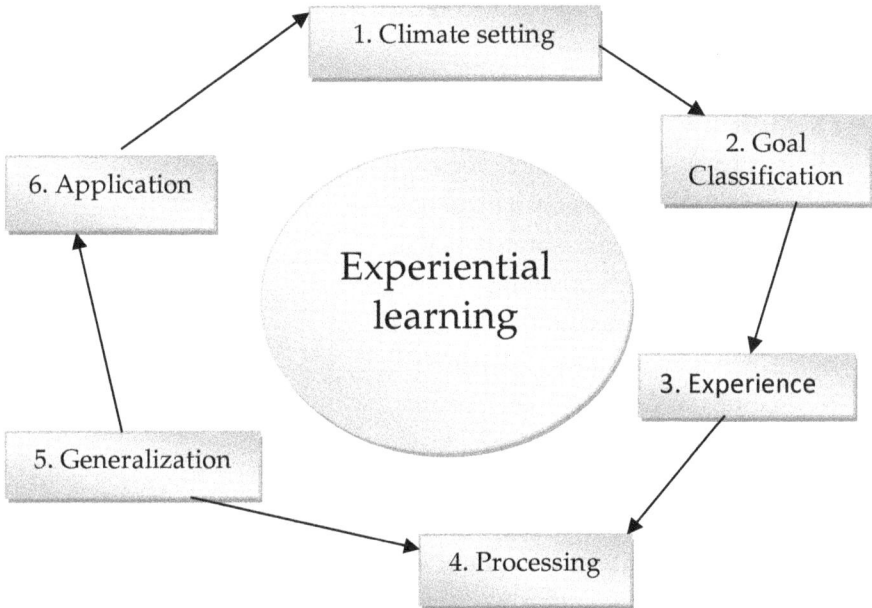

Figure 11: Experiential Learning

Steps are:

1. Climate setting
2. Goal classification
3. Experience
4. Processing
5. Generalization
6. Application

All these steps can be measured to find out gaps against the standard elements.

Training Areas of Different Subjects

1. Horticulture

1. Flower and fruit drops
2. Cultivation practices of lemon

3. Fruit crops like banana, pine apple, papaya cultivation practices
4. Pointed gourd
5. River bed cultivation of potato and water melon
6. Cultivation and processing of cinnamon
7. Fruit and flower drops in mango
8. Management of kharif and rabi vegetables
9. Spine gourd commercial cultivation
10. Packages for summer vegetables
11. Vegetable seed production packages
12. Cultivation of hybrid cabbage and cultivation
13. Micronutrients use in vegetables
14. Post harvest technology of vegetables and fruits
15. Care and management of nurseries
16. Packages for tissue culture banana
17. Hybrid Tomato cultivation
18. Hybrid brinjal and chili cultivation
19. Cultivation packages of potato
20. Planting techniques of fruit plants
21. Off-season vegetable cultivation
22. Harvesting and storage of ginger and turmeric
23. Packages of practices for garlic and onion
24. Planting techniques of turmeric and ginger
25. Commercial cultivation of cashew nut
26. Care and management of orchard
27. Budding of marigold and tuber rose
28. Cultivation of seasonal flowers
29. Management of floriculture nurseries

2. Plant Protection

1. Stem borer control in paddy
2. B.P.H. control of paddy
3. Integrated disease management in paddy
4. IPM in cotton
5. IPM in ground nut
6. Integrated disease management in betlevine
7. Pest and disease control in turmeric and ginger
8. Fruit borer control in tomato
9. Pest management in winter vegetables

10. Shoot and fruit borer in brinjal

11. Insect and pest control in rape seed mustard

12. IPM in red gram

13. IPM in sugar cane

14. Integrated pest management in paddy

15. Pest and disease control in summer vegetables

16. IPM in citrus

17. Insect control in mango

18. Insect and disease management in banana

19. Common diseases of vegetables and their control measures

20. Safe storage of food grains

21. Bee keeping for self-employment

22. Mushroom cultivation

23. Mushroom spawn production for self employment

24. Mass production of bio-control agents

25. Virus disease of crops and their control measures

26. Use of plant production for pest control

27. Preparation and application of spray solutions

28. Rodent control

3. Fishery

1. Fresh water pearl production

2. E.U.S. Fish disease and it's prevention

3. Pre-stocking management in crap culture

4. Prawn in crap poly-culture

5. Common fish diseases and their preventive measures

6. Preservation of fish

7. Composite fish culture

8. Crap seed production

9. Income generation through fingerling raising

10. Paddy-cum-fish culture

4. Extension Education

1. Extension training methods and their application

2. Techniques of conducting successful demonstration

3. Techniques of developing agro-entrepreneurship among rural women

4. Methodology in project management

5. Techniques of conducting training on ELC model

    6. Use of PRA-Techniques

    7. Preparation of simple aids to make training effective

    8. Skills in use of motivational techniques

    9. Identification of adopter category to accelerate rate of adoption

    10. Techniques for working with farm women

5. Agriculture Engineering

    1. Use of power tillers and it's matching equipments

    2. Use of paddy thresher and winnowers

    3. Use of manually operated weeders

    4. Use of bullock drawn puddler for low land rice cultivation

    5. Use of poser reaper

    6. Operation, adjustment and use of multi crop seed drill

    7. Use of groundnut decorticator

    8. Use of groundnut threshers

    9. Use of manually operated paddy transplanter and power transplanter

    10. Making hullers modernized

    11. Use and maintenance of pedal operated sunflower thresher

    12. Use and repair of mould board iron plough for land preparation

    13. Use, operation and maintenance of power tiller and matching equipments

    14. Use and repair of small handy implements and tools

    15. Use of low lift pumps for irrigation

    16. Use and repair of diesel pumps

    17. Soil moisture conservation measures

6. Field Crops (Agronomy)

    1. Use of BGA and Azolla for transplanted rice

    2. Use of Azosprillum and Azotobactor in rice

    3. Use of PSB and potash culture in rice

    4. Use of Rhizobium culture in rice

    5. Nursery management and transplanting techniques in paddy

    6. Raising of green manure crops in rice

    7. Fertilizer management in HYV and hybrid rice

    8. Making of good compost pits

    9. Techniques of soil sample collection for testing

    10. Making of vermin compost

    11. Nitrogen management in rice

    12. Weed control techniques in paddy

13. Management of acid soils
14. Fertilizer management in summer groundnut
15. Niger cultivation in degraded soil
16. Cuscuta control in Niger
17. Toria as a catch crop between two rice crops
18. Intercultural practices in sunflower
19. Sunflower cultivation
20. Sub-plotting and sowing of wheat
21. Planting techniques of sugarcane
22. Alternate land use system
23. Integrated nutrient management in sugarcane
24. Water management in wheat
25. Poly mulching in groundnut
26. Intercropping of paddy with arhar
27. Inter cropping of maize with cowpea
28. In situ moisture conservation by trenching methods
29. Organic farming in paddy
30. Seed farming in paddy
31. Seed production techniques in cereals

7. Home Science

1. Preservation of locally available fruits
2. Value addition to locally grown fruits and vegetables
3. Appliqué work for rural women
4. Soft toy making for income generation
5. Supplementary diet preparation for pre-school children
6. Preparation of potato chips, rice papad, biripapad etc.
7. Knitting for rural women
8. Fabric painting
9. Storage of cereals and pulses
10. Seed extraction in tomato
11. Seed treatment
12. Nutritional garden
13. Agarbati making
14. Making of simple garments
15. Preparation of nutritional diets
16. Processing and value addition to turmeric
17. Preparation of ginger pickles and squash

18. Mushroom cultivation
19. Use of tools for reducing drudgery
20. Jute craft for income generation
21. Fabric printing by block printing
22. Preparation of decorative by golden grass

(The training needs of farmers of different climatic zones of the state were collected by different KVKs of OUAT, Bhubaneswar)

Before starting evaluation of training program, there is need to know all the elements involved in the training process. Extension Institute of Hyderabad has included as much as 11 elements of training programme. These are:

1. Training needs
2. Training results
3. Training policy
4. Training objectives
5. Training audience
6. Training program
7. Training staff
8. Training facility
9. Training budget
10. Training time
11. Training evaluation

All these elements can be measured in terms of extent to which they are suitably framed/arranged.

Measurement of training programmes aiming at capacity building should cover the aspects like: (i) Master training plan, (ii) Training proposal, (iii) Development of training curriculum, (iv) Training process, (v) Training need assessment, (vi) Training objectives, (vii) Training methods, (viii) Lesson plan, (ix) Evaluation. All these items can be measured in terms of quantitative and qualitative form.

Impact Evaluation of Training Programme

The impact evaluation of training programme implies that what difference training made with the participants after training is over. Different methods are applied to measure the impact or difference. The questions are by many evaluators that what aspects of training output are to be assessed. The impact evaluation of training programme may be classified into three aspects.

1. Learning impact
2. Performance impact
3. Field Impact

The focus of training impact assessment can be illustrated indicating training, effects and impact components.

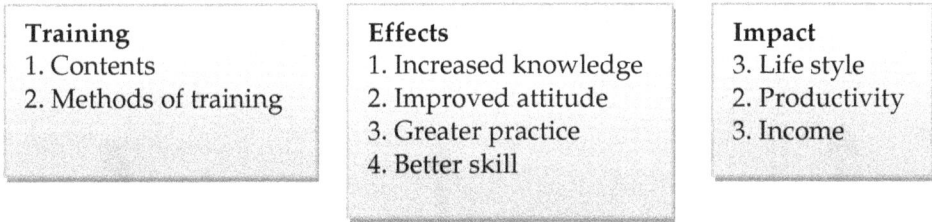

Training	Effects	Impact
1. Contents	1. Increased knowledge	3. Life style
2. Methods of training	2. Improved attitude	2. Productivity
	3. Greater practice	3. Income
	4. Better skill	

Figure 12: Training Programme

Assessment of Impact

The training programme deals with living organism *i.e.* Human beings. When an individual is exposed to training, an interaction among five elements takes place. These are: (i) Trainer, (ii) Learner or participant, (iii) Material of instruction, (iv) Physical facilities and (v) Subject matter. Therefore to evaluate training programme one has to evaluate all the five elements to draw a conclusion about impact of training.

The analytical assessment of a training programme covers all the elements of teaching and learning process which can be represented as:

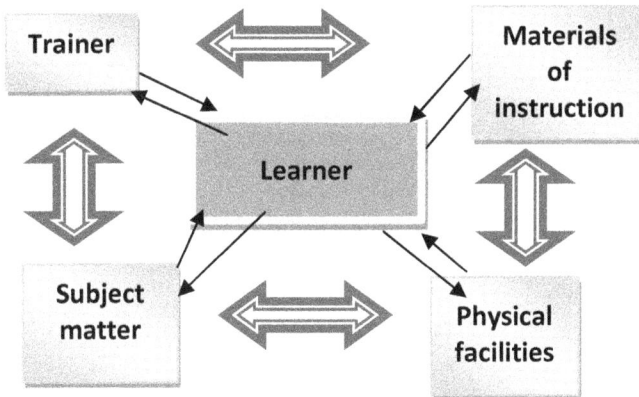

Figure 13: Teaching and Learning Process.

An evaluation plan needs to be developed to measure different aspects of training with a specified format so that evaluation becomes easy and perfect.

Measurement of Knowledge Gain

Training is an educational activity. At the end of a training programme we are interested to know what trainees have learned and how much is the retention and how much are put to practice. In other words, what is impact of training?

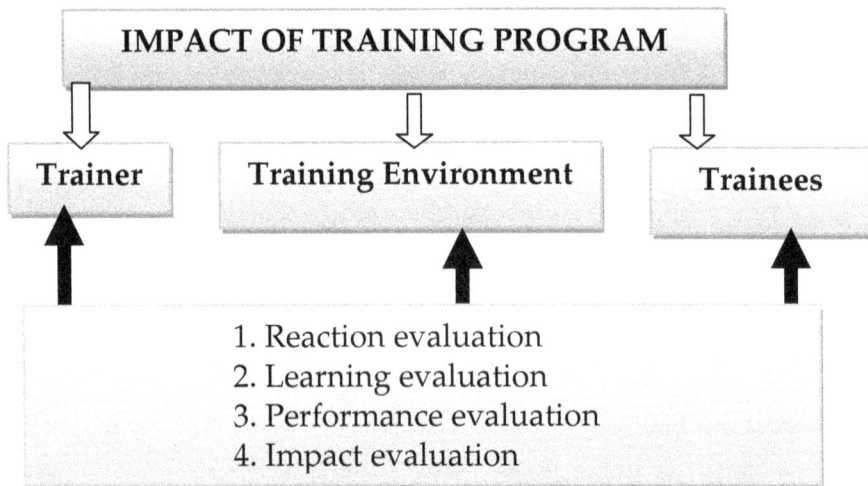

Figure 14: Impact of Training Programme.

Table 21: Evaluation Frame Work

Areas of Training for Evaluation	Parameters to be Measured	Methods/Tools to be Used
1.Training Programme	1. Need assessment 2. Training objectives 3. Content of training 4. Presentation of content 5. Venue for training 6. Training time 7. Use of A.V. aids	
2. Knowledge	1. Knowledge of facts 2. Knowledge of practice 3. Knowledge of method 4. Knowledge of subject	
3.Skill	1. Skill of practice 2. Skill of application 3. Skill of demonstration 4. Skill of problem solving 5. Skill of evaluation 6. Skill of communication	
4. Attitude	1. Attitude towards self 2. Attitude towards practice 3. Attitude towards learning	
5. Behaviour	1. Covert behaviour 2. Overt behaviour	
6.Competency	1. Technical competency 2. Professional competency 3. Economic competency	

The impact assessment follows statistical treatment. To determine the impact of training we shall have to adopt research design. The research design follows two approaches. One is experimental design and other is ex-post-facto. In case of experimental design, the cause-effect relationship is worked out by manipulating independent variables *i.e.*, methods of training, time, trainers, subject matter etc. are stretched as per requirement to find their relative effect, *i.e.*, learning. In case of ex-post facto approach we start from effect to ascertain the causes and their dimensions. Training when imparted and impact is assessed it is the experimental approach.

Experimental Research Design

Design is the plan, structure and strategy of the training in which we are interested to include in experimental design. With training programme taking care of trainer, trainees, subject matter, materials of instruction and physical facilities, we have to plan for experimentation.

- ☆ Step 1. When we conduct training, and training is over, we want to measure effect or outcomes.

- ☆ Step 2. We decide appropriate statistical treatment to measure the effect.

- ☆ Step 3. Make interpretation of result concluding extent of gain in knowledge, skill and attitude.

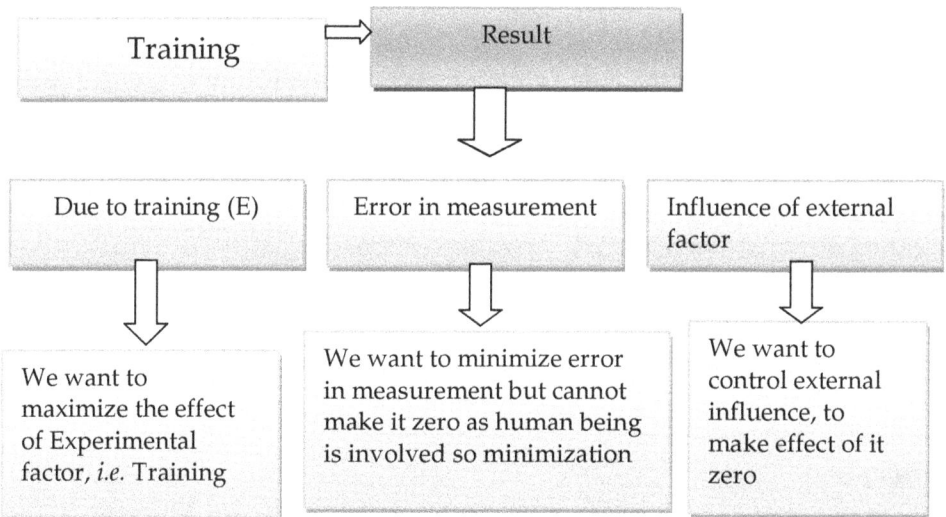

Figure 15: Training Result.

Under such cases we can measure the effect of training as illustrated.

Experimental value = Minimum human error + control of external factor. It is called **MaxMinCon,** means maximization of experimental variable, minimization of error variable and control of external variable.

How to Do It?

There are two ways of doing such experiment. One is control group approach and another is Before-After design.

1. Control Group Approach

We can take two homogeneous groups. In one group we shall impart training (Experimental E) and not in other group called Control group (C).

Experimental group E	Control group C

Gain in knowledge due to training = E-C (Impact)

Problem

1. Difficult to get homogeneous group
2. Difficult to keep human beings in control condition
3. Difficult to keep trained people from untrained people

To overcome the problem we have to opt for Before-After design.

2. Before-After Design

In case of Before-After design there is no need of taking control group. We can take Pre-measurement (Pre-M) of the experimental group before starting training and Post-measurement (Post-M) after training is over to assess impact of training. The difference between two measurements will give the extent of impact due to training (experimental variable)

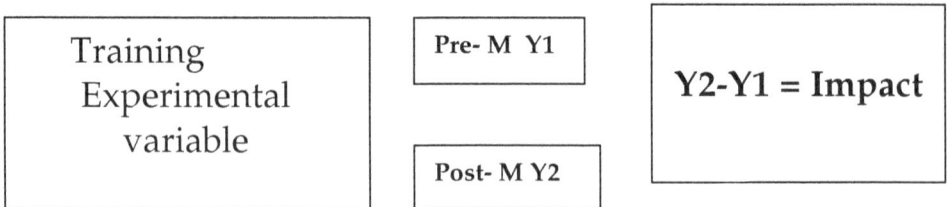

Training Experimental variable	Pre- M Y_1 Post- M Y_2	$Y_2 - Y_1$ = Impact

Example

In a training programme, the trainer offered training to women of SHG groups for two days covering horticultural crops. The scoring was adopted on a three point continuum like, learned as expected, Learned to some extent, Learned very little with assigned scores of 3, 2 and 1.

Table 22: Before-After Design

Sl.No.	Training Content	Learned as Expected	Learned to some Extent	Learned Little
1.	Vegetable cultivation			
2.	Nursery management			
3.	Plant Protection in mango			
4.	Vegetable preservation			
5.	Seed production			
6.	Commercial horticulture			

1. Maximum score will be: 18

2. Minimum score will be: 6

3. Group size: 5

Table 23: Average Score in Pre and Post-Measurement

Sample	Pre-M	Post -M
R1	12	16
R2	14	18
R3	10	12
R4	12	14
R5	10	12
Total	58	72
Average =11.60		14.4
Per cent in gain knowledge 24.13		

This 24.13 per cent result is due to: (i) Experimental variable (training) + Error in measurement+ variation due to external variable. Applying Max Min Con concept we have to minimize error due to measurement and influence of external variable

Reasons behind **Max Min Con**

Suppose total Variance   VT

Variance due to training (experimental variance)   Vt

Variance due to error in measurement   VE

Variance due to external factor   Ve

Equation VT= Vt + VE + Ve

If VE and Ve are made zero, then VT=Vt that we want to achieve. As VE is human error it cannot be zero but minimized and Ve due to external factor can be controlled.

Methods of Minimization of (VE) Error due to Measurement

1. Non response error of the respondent is to be avoided.
2. Avoid response bias data recording.
3. Care of instrument associate errors like poor questions, wrong wording and ambiguous complex are to be avoided.
4. Situation error. This occurs when third person is present the respondent tries to hide some facts. This has to be checked.
5. Measurement as source of error to be checked.

Control of External Variable

External variable is unwanted in experiment. The variable is to be controlled. There are five possible ways to control external variable.

1. *To eliminate variable as variable*: When the sample is homogeneous the effects of external variable can be eliminated. It is to make desired trait equal at the starting of experiment. If all the trainees are equal in qualification at basic level and external the effect of difference in qualification can be eliminated.
2. *Use of randomization techniques*: When the sample for response is selected following random sampling the effect of external factors get equally distributed and as such effect of it is eliminated.
3. *To match subject*: When a matching variable is substantially correlated on dependent variable, we should do paring on equal traits like age, education caste etc. to eliminate effect of external variable.
4. *Statistical control*: There are many statistical methods which can successfully control the effect of external variable. The design like, completely randomized design, randomized block designs can be used for the purpose.
5. Build external variable right in design so that external variable is controlled and additional information generated for external variable can be otherwise used.

In experimental design to get control situation is very difficult, therefore the experimental variable should be stretched as far as possible. In other words, the difference between treatments should be very distinctive. Example: E1 = Primary Education E2 = Middle school education E3 = High school education etc. The difference should be clear cut between primary and middle school by cutting point. Education up to class V is primary and from Class VI it is middle school education.

Chapter 5

Measurement of Adoption: Behaviour and Innovativeness of Farm Women

Adoption and innovativeness are closely related. The adoption as a concept reveals that an individual has to translate information or innovation or technology into action and derive satisfaction out of it. To achieve this or to be called as adopter one has to undergo different stages. The different stages of adoption process have been postulated by authors differently with variation in stage concept. The variation in stage concept is due difference in observation. The commonly accepted stages in adoption process are:

1. Awareness
2. Interest
3. Evaluation
4. Trial
5. Adoption

Besides, the additional steps suggested by different authors are:

1. Conviction
2. Acceptance or complete adoption
3. Desire
4. Action
5. Satisfaction
6. Decision-making

7. Information

8. Need

9. Deliberation

10. Integration

11. Knowledge

12. Persuasion

13. Decision

14. Confirmation

All these stages are recognized in relation to theories of learning and teaching. To avoid confusion about the stages of adoption it is better to go with commonly accepted stages like, awareness, interest, evaluation, trail and adoption. All the stages are interrelated, but not rigid and individuals may skip some stages depending on situation, type of technology and social system.

What and How to Measure Adoption Behaviour?

Measurement of adoption behaviour can be considered from two aspects. These are, quantitative and qualitative aspects. In social research qualitative measurement has special significance while quantitative measurement is essentially required to arrive at conclusions.

Adoption process brings changes internally and externally in adopters may be male or female. The covert and overt behaviour provides scope to measure the qualitative aspect in quantitative terms. The covert and overt behaviour observed with adopters are summarized and given herewith.

Table 24: Covert Behaviour and Overt Behaviour

Stages of Adoption Process	Covert Behaviour	Overt Behaviour
1. Need	Arousal	1. Readiness to change 2. Dissatisfaction with present situation 3. Expression of feeling occasionally 4. Seek mass media for information
2. Awareness	Psychological process, selecting stimulus and perception	1. Seeing and hearing of innovation 2. Feeling towards possible solution 3. Mind for expressing search for solution 4. Seek mass media for information 5. Consult with informal sources
3. Interest	Orientation and exploration	1. Gathering of information 2. Cares for those who talks of it 3. Look for mass media 4. Seek literature 5. Tends to contact resource person
4. Deliberation	Community of reorganization	1. Diffusion of own feeling 2. Tries to find those are in similar problem 3. Weighs the problem on own situation

Contd...

Table 24–*Contd...*

Stages of Adoption Process	Covert Behaviour	Overt Behaviour
5. Evaluation	Reinforcement of learned innovation	1. Collection of further information 2. Use of formal and informal sources of information 3. Consideration of own affordability 4. Analysis of risk 5. Close consultation with family members 6. Looking of availability of inputs 7. Considering market value and demand 8. Consideration of subsidies
6. Trial	Expectancy	1. Tentative adoption 2. Doubts in mind 3. Looking after other adopters 4. Contact with progressive farm women/farmer 5. Enquire from neighbours, relatives 6. Contact with extension agencies
7. Adoption	Acquiring new drives	1. Builds trust on technology 2. Make standard use of technology 3. Looks for refinement and modification 4. Discuss about market needs 5. Calculates loss/profit 6. Thinks how to reduce cost of production 7. Talks success to others 8. Plan to integrate in own system 9. Decides points of success/failure 10. Decides for expansion under technology
8. Integration	Generalization of learned skill, consumentary response	1. Change in attitude 2. Change in skill 3. Change in knowledge 4. Strong supporter of technology 5. Advise others about technology 6. Derives satisfaction 7. Looks for more refinement 8. Brainstorming on technology improvement

The contents of above summary reveal the visible systems of an individual passing the stages of adoption process. Basing on the visible symptoms, adoption behaviour can be measured applying three point continuum.

Table 25: Adoption Stage 'NEED' (Symptoms) N=150

Sl.No.	Symptoms	Very Often	Often	Sometimes	Average Score
1.	Readiness for change	70	30	50	1.67
2.	Dissatisfaction with present practice	55	72	23	2.21
3.	Attitude for diversification	68	35	47	2.14
4.	Full of questions in mind	74	39	37	2.24
5.	Searching after successful farmers	38	67	45	1.95

A study conducted in Odisha by Satapathy and Mishra in the context of adoption behaviour of SHG members (2008) is illustrated herewith to indicate measurement of symptoms of adoption stages.

Table 26: Adoption Stage 'AWARENESS' (Symptoms)

Sl.No.	Symptoms	Very Much	Much	Little	Average Score
1.	Seeing and hearing of innovation	120	20	10	2.73
2.	Contact with media	40	80	30	2.06
3.	Contact with progressive persons	52	46	52	2.00
4.	More collection of information	28	42	80	1.65
5.	Comparing own situations with others	19	62	69	1.67

Table 27: Adoption Stage 'DELIBERATION' (Symptoms)

Sl.No.	Symptoms	Very Much	Much	Little	Average Score
1.	Expression of problems before others	30	28	92	1.58
2.	Looking for related events	41	31	78	1.75
3.	Interaction with technical people	17	18	115	1.86
4.	Searching of similar problem facing persons	12	16	122	1.34
5.	Seriously looking for way out	26	28	96	1.53

Table 28: Adoption Stage 'INTEREST' (Symptoms)

Sl.No.	Symptoms	Very Frequently	Frequently	Occasionally	Average Score
1.	Gathering of detail information	72	30	48	2.16
2.	Search for details of practice	42	37	71	1.81
3.	Tempted to visit successful sites/events	28	27	95	1.55
4.	Interaction with experts	16	22	112	1.36
5.	Visit of demonstration/exhibition	12	23	115	1.31

Table 29: Adoption Stage 'TRIAL' (Symptoms)

Sl.No.	Symptoms	Very Much	Much	Little	Average Score
1.	Trial of technology with care	40	49	61	1.86
2.	Tends for clarification of doubts	32	59	59	1.82
3.	Careful observation o key points	30	65	55	1.83
4.	Close observation of performance	20	48	80	1.59
5.	Look for own deficiencies	38	32		1.72

Table 30: Adoption Stage 'EVALUATION' (Symptoms)

Sl.No.	Symptoms	Very Frequently	Frequently	Sometimes	Average Score
1.	Compare with existing practice	38	42	70	1.78
2.	Judging own affordability	354	40	75	1.74
3.	Examination of possible profit	29	39	82	1.65
4.	Consultation with earlier adopters	17	23	110	1.38
5.	Weighing of market demand and outputs	36	49	65	1.81

Table 31: Adoption Stage 'ADOPTION' (Symptoms)

Sl.No.	Symptoms	Very Frequently	Frequently	Sometimes	Average Score
1.	Increase in area/unit	38	32	30	1.38
2.	Increase in profit	35	46	69	1.77
3.	Increase in satisfaction	30	57	63	1.78
4.	Comparing own success with others	19	63	66	1.66
5.	Contact with experts	16	28	106	1.38

Table 32: Adoption Stage 'INTEGRATION' (Symptoms)

Sl.No.	Symptoms	Very Much	Much	Little	Average Score
1.	Change in knowledge, skill and attitude	38	49	63	1.83
2.	Makes practice regular	30	36	84	1.64
3.	Includes in annual programme	28	32	116	1.74
4.	Adivises others about the practice	17	19	114	1.35
5.	Interest for commercialization	15	35	100	1.43

A look at the scores obtained by sample in all the stages of adoption process is:

Table 33: Stages of Adoption Process

Sl.No.	Stages	Average Score	Sl.No.	Stages	Average Score
1.	Need	2.04	2.	Awareness	2.02
3.	Deliberation	1.61	4.	Interest	1.64
5.	Trial	1.76	6.	Evaluation	1.67
7.	Adoption	1.59	8.	Integration	1.59

Table 34: Women in Adoption Behaviour

Sl.No.	Stages of Adoption Process	Maximum Obtainable Score	Minimum Obtainable Score
1.	Need	15	5
2.	Awareness	13	5
3.	Deliberation	15	5
4.	Interest	15	5
5.	Trial	15	5
6.	Evaluation	15	5
7.	Adoption	15	5
8.	Integration	15	5
	Total	120	40

But when we compare the movement of men and women adopters in the stages of adoption we find women are better adopters.

Category	Range of Score
1 High adoption	58 and above
2. Medium adoption	50-57
3. Low adoption	Up to 49

The data are indicative of facts that movement of women towards adoption and integration is higher than men. It is a fact that once women are convinced of a new practice they adopt at higher percentages than their counterparts.

Table 35: Movement of Men and Women at different Stages of Adoption Process (per cent)

Men	Stages of Adoption Process	Women
100.00	1. Need	100.00
85.00	2. Awareness	65.00
70.00	3. Deliberation	48.00
55.00	4. Interest	38.00
32.00	5. Trial	30.00
24.00	6. Evaluation	28.00
8.00	7. Adoption	14.00
3.00	8. Integration	9.00

Very often the question of adoption is associated with area spread and number of persons adopting the technology. The two schools of thought are genuine. The area under new technology of an individual is a good indicator of adoption while other side is that out of the population of a village or community how many have adopted the technology irrespective of area spread. From learning point of view, more the persons are adopting the technology better is the level of achievement. Even though number of adopters is more but the area spread is less, then the national target remains unfulfilled.

These two arguments need consideration depending on objectives, programmes and aims of projects.

Adopter Categories

In the domain of adoption studies, the concept of "Adopter Category" has special position as it indicates individuals in adoption scale. In simple sense it reveals who takes how much time to adopt a practice or technology. The concept is based on hypothesis rather truth that all individuals take their time to accept and adopt a practice. Some are quick whereas some are very late to adopt an innovation.

Innovativeness is the degree to which an individual is earlier to adopt the practice than the members of his social system. On the basis of innovativeness we classify adopters into different categories.

Table 36: Classification of Adopters

Sl.No.	Adopter Category	Percentage
1.	Innovators	2.59
2	Early adopter	13.50
3.	Early majority	34.00
4.	Late majority	34.00
5.	Laggards	16.50

Table 37: Measurement of Innovativeness of Individuals in Adopter Category

Sl.No.	Characteristics of Innovators	Degree to which Individuals can be Classified			
		Very Much (4)	Much (3)	Little (2)	Not at All (1)
1.	Venturesome, adopt immediately without much verification				
2.	Always eager to try new practice/ idea				
3.	Always try to be of the local group and keeps more contact outside				
4.	Take risk without much analysis				
5.	Acquainted with facing of uncertainty				

Table 38: Characteristics of Early Adopters

Sl.No.	Characteristics of Early Adopters	Degree to which Individuals can be Classified			
		Very Much (4)	Much (3)	Little (2)	Not at All (1)
1.	Commands respect in society				
2.	Have regards for local system				
3.	Opinion are sought in many matters				

Contd...

Table 38–*Contd...*

Sl.No.	Characteristics of Early Adopters	Degree to which Individuals can be Classified			
		Very Much (4)	*Much (3)*	*Little (2)*	*Not at All (1)*
4.	Perfect checking of practice before adoption				
5.	Contact with extension agency				
6.	Role model in village				
7.	Educational status				
8.	Possess of land				
9.	Comfort in income level				
10.	Quickness to adopt ideas in own situation				

Table 39: Characteristics of Early Majority

Sl.No.	Characteristics of Early Majority	Degree to which Individuals can be Classified			
		Very Much (4)	*Much (3)*	*Little (2)*	*Not at All (1)*
1.	Long discussion on topic before adoption				
2.	Earliness to adopt new practice				
3.	Contact with local source of information				
4.	Leadership style				
5.	Quickness in taking decision				
6.	Maintaining of neutral position in decision				
7.	Looking at success of others before final decision				
8.	Calculation of loss and profit				

Table 40: Characteristics of Late Majority

Sl.No.	Characteristics of Late Majority	Degree to which Individuals can be Classified			
		Very Much (4)	*Much (3)*	*Little (2)*	*Not at All (1)*
1.	Degree of investment				
2.	Look, see and adopt practice				
3.	Economic compulsion to accept new practice				
4.	Need pressure to accept new practice				

Contd...

Table 40–*Contd...*

Sl.No.	Characteristics of Late Majority	Degree to which Individuals can be Classified			
		Very Much (4)	Much (3)	Little (2)	Not at All (1)
5.	Education back ground				
6.	Income flow				
7.	Mobility outside locality				
8.	Contact with extension agency				
9.	Leadership position				

Table 41: Characteristics of Laggards

Sl.No.	Characteristics of Laggards	Degree to which Individuals can be Classified			
		Very Much (4)	Much (3)	Little (2)	Not at All (1)
1.	Traditionalism				
2.	Leadership position				
3.	Social mobility				
4.	Information seeking habit				
5.	Idea about past				
6.	Decision on past experience				
7.	Followership role				

Factors of Adoption

Very often we are required to measure the factors of adoption of technology. This part of measurement requires quantitative aspects rather qualitative one. Vast volumes of adoption and diffusion studies have found out that the factors of adoption can be grouped as follows.

1. Personal factors
2. Social factors
3. Economic factors
4. Cultural factors
5. Technological factors
6. Environmental factors
7. Psychological factors
8. Extension factors
9. Policy factors
10. Input supply factors
11. Marketing factors.

Chapter 6

Measurement of Communication Behaviour of Farm Women

Communication is the day to day affair of each and every body. Communication behaviour of an individual is influenced by a multiple factors. The factors like source, receiver, channel, message and socio-cultural situations. The communication domain of rural women is family members and household activities, social interaction with neighbour, village, community and society at large. Our interest is that how rural women communicate with extension personnel from whom they receive technical information and transmit them in family and outside family. It is known that more the communication better is the exposure and higher is rate of gaining information.

In communication process, the sender of source interacts with receiver, channel and message and produces feedback. The feedback contains reaction of the receiver which provides the ideas about extent of absorption of information or gain in knowledge. The farm women when receives technical massage from any sources may be formal or informal she becomes recipient of message or receiver. She becomes source of message when she transmits it to the family members, neighbours, etc. The interaction of women with sources, receiver, message and channel together gives the ideas of her communication behaviour. The aim is to measure the

Average Communication Pattern of Rural Women per Day		
Sl.No.	Items	Per cent of Time
1.	Children	16.00
2.	Husband	28.00
3.	Other members of family	12.00
4.	Livelihood activities	32.00
5.	Neighbors	9.00
6.	Outsiders	3.00

communication behaviour in terms of positive or negative which decides rate of transfer of knowledge. The chapter is meant for the women who assume the responsibility of leading the group and involved in group dynamics. It is mostly applicable to women key communicators or influential.

Women as Source of Communication

We now emphasize empowerment of women. To achieve this goal, there is need to increase communication ability of women. The communication ability or traits are measurable. Both qualitative and quantitative methods of measurement can be applied to measure communication behaviour.

At present the development programmes for women expect that rural women should understand the technology, adopt it and spread the message in society for mass adoption. The desirable characteristics or traits that women should possess are,

1. Ability to initiate purposeful message
2. Directs message to her audience
3. Have faith that her message will be accepted
4. Communication should be friendly
5. Knows well the fate of communication

Indicators for Measurement

Measurement of communication ability can be dealt with: (i) expression ability, (ii) message of communication,(iii) selection of channel and (iv) treatment of message.

1. Expression Ability

The communication domain of rural women is group oriented.

At one time, her audience would be around 10-15 only. The degree to which she can express her technical ideas to the members of the group can be taken as her expression ability.

Communication ability=Ratio between message thought to communicate and actual communication of the message.

Suppose one thinks to compose message with ten important points and expressed only eight points. Ratio is 8/10 or 80 per cent communication ability.

Communication ability of one is cumulative effects of: (i) purposeful initiation of message, (ii) characteristics of message she delivers, (iii) type of channel she chooses and (iv) type of treatments she gives to message.

2. Credibility of Communicator

Leagans (1963) defined credibility of communicator as perceived by audience is a powerful determinant in communication process. According to him a good communicator should have the following characteristics to make his/her communication effective.

1. Knowledge about	**2. Have interest in**
1. Objectives	1. Welfare of audience
2. Audience	2. How to reach audience
3. Message	3. Result of communication
4. Channel	4. Use of channel
5. Ability and limitation	5. Process of communication

Credibility of Communicator

3. Must Prepare	**4. Possess Skill**
1. Plan to communicate	1. Selection of message
2. Communication material	2. Treatment of message
3. Methods of evaluation	3. Expression
4. Feedback collection	4. Selection of channel
5. Good ground to communicate	5. Understanding audience
	6. Assessment of result

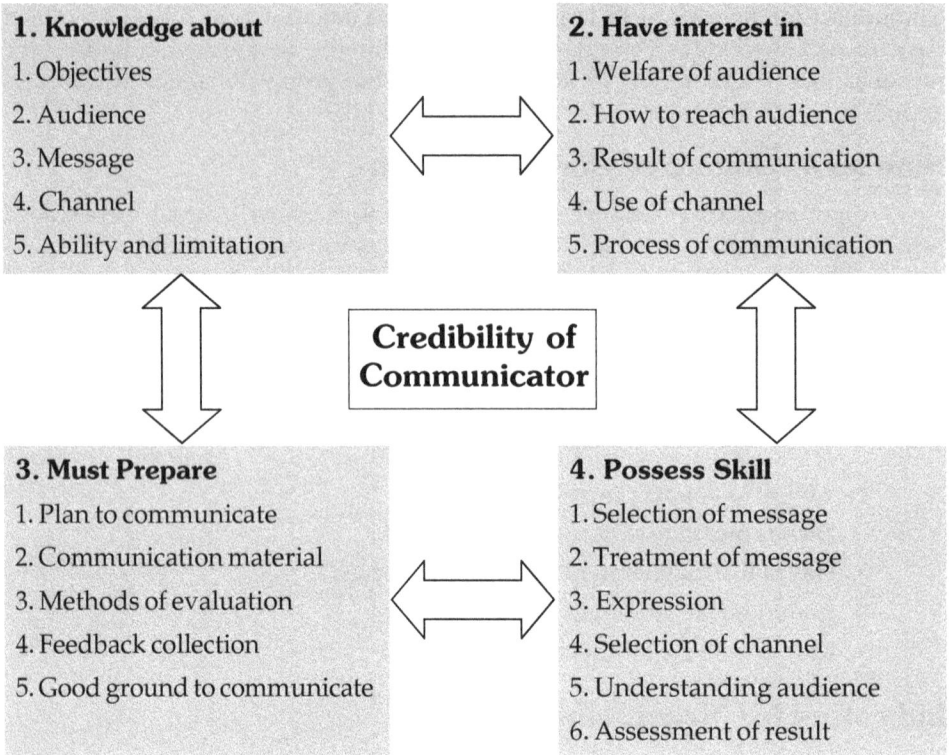

Measurement of Credibility of Communicator

The attributes of credibility can be measured in a three point continuum to know the degree to which these are possessed by the communicators.

Attributes	Very Much	Much	Little
1. Knowledge about			
1. Objectives			
2. Audience			
3. Message			
4. Channel			
5. Ability and limitation			
2. Interest in			
1. Welfare of audience			
2. How to reach audience			
3. Use of channel			
4. Process of communication			

Contd...

Attributes	Very Much	Much	Little
3. Must prepare			
1. Plan of communication			
2. Communication material			
3. Methods of evaluation			
4. Plan for feedback collation			
5. Ground for effective communication			
4. Possession of skill			
1. Selection of message			
2. Treatment of message			
3. Expression			
4. Channel selection			
5. Understanding audience			
6. Assessment of result of communication			

Maximum score will be = 60   Minimum score will be = 20

	Category	Score Range
1.	High communication behaviour	44 and above
2.	Medium communication behaviour	33-43
3.	Low communication behaviour	Up to 32

3. Characteristics of Message

For effective communication that too in rural areas, women communicator has to carefully look into the characteristics of messages that bring success or become effective. The messages beamed by the women communicators can be checked with desirable and effective traits. Taking sample messages of women communicators may be minimum of 10; one can calculate the presence of desirable traits in the message. For example, the procedure followed in a study conducted at Odisha 2008 with a sample of 25 key communicators is explained herewith (N = 25).

Sl.No.	Presence of Desirable Traits	Very much Present	Present to some Extent	Little or Not Present	Average Score	Gap (per cent)
1.	Convincing	12	10	2	2.32	22.67
2.	Need based	10	12	3	2.36	21.13
3.	Clear understanding	8	12	5	2.12	29.33
4.	Accurate	15	5	5	2.00	33.33
5.	Timely	7	12	6	2.04	32.00

Contd...

Sl.No.	Presence of Desirable Traits	Very much Present	Present to some Extent	Little or Not Present	Average Score	Gap (per cent)
6.	Complete	6	8	11	1.80	40.00
7.	Simple	9	10	6	2.12	29.33
8.	Attention catching	6	6	13	1.72	42.67
9.	Newness	7	7	11	1.84	27.00
10.	Practicable	12	8	5	2.28	24.00
11.	Affordable	15	9	1	2.56	14.67
12.	Immediate problem solving	11	13	1	2.40	20.00
13.	Tunes with capability	11	6	9	2.16	28.00
14.	Satisfactory	5	8	12	1.72	42.67
	Average	9.57	8.85	6.43	2.10	30.00

The score may be assigned as 3, 2, and 1 for very much, to some extent and little or not present. Applying same procedure we can classify them into three categories and find out gaps to which the communicator has to give more emphasis.

(i) Categorization

To classify the respondents on communication behaviour (traits), we can follow the range of score out of maximum of 42 and minimum of 14.

	Category	Score Range
1.	High communication level	31 and above
2.	Medium communication level	23-30
3.	Low communication level	22 and less

(ii) Gap Analysis

To find out gap in communication we have to take

Maximum obtainable score in each case =3 and actual score obtained against each item expressed in percentage.

$$\text{Gap (\%)} = \frac{\text{Maximum Obtainable Score} - \text{Actual score obtained}}{\text{Actual Score Obtained}}$$

Item No. 1. $3.0-2.32/3 \times 100 = 22.6$

1. Treatment of Message

Treatment of message is the core quality of the communicator. He/She is the best communicator who puts content and language (code) into a palatable form for the receiver. The skill of communicator lies with as how to prepare and produce good messages. In treatment of message, the content remains unchanged but shape is given to language for better acceptance of message. Measurement of this part of the communication process is relatively difficult. From opinion of the audience (feedback) the quality of treatment of message can be ascertained. Determination of quality of treatment depends on the following indicators:

1. Reaching of desired target group
2. Treatment is not same for all messages
3. Message is to be attractive
4. Emotion in message is to be reflected
5. Treatment to indicate the desire of the communicator
6. Content to remain unchanged
7. Who, why, how, where whom and when should be indicated in action message

Sl.No.	Statements	Score Encircled			
1.	Reaching of desired target group	4	3	2	1
2.	Attractiveness	4	3	2	1
3.	Reflection of emotions	4	3	2	1
4.	Focus of the communicator	4	3	2	1
5.	Reflection of 5 Ws	4	3	2	1
6.	Unchanged content	4	3	2	1
7.	Suitability of message	4	3	2	1
8.	Selection and use of appropriate words	4	3	2	1
9.	Sequence of items	4	3	2	1
10.	Coverage of subject matter	4	3	2	1

Comparing the opinions of the receivers about the treatment of message conclusions can be drawn about its effectiveness.

The statements qualifying treatment of message can be rated as 4 for very much, 3 for much, 2 for little and 1 for none.

The scores obtained by the respondents against each statement and their corresponding average will provide a clear cut picture as to what extent the treatment of message is effective.

Channel Selection and Use

Women communicator when want to send the message need to identify appropriate channel. In case of SHG, small group and village meeting the women mostly use verbal communication. The selection of channel depends upon certain important factors for effective communication. These are: (i) What channels are available in rural areas, (ii) Preference of source for the channel, (iii) Adaptability of channel to the kind of messages, (iv) Impact of channel, (v) Cost involvement,

(vi) Compatibility with content of message and (vii) Preference along with availability of channel with receivers.

A good women communicator is interested in acceptance and adoption of the message by her receivers. To achieve this goal, the factors that regulate the acceptance and reasons for resistance are to be known. These are:

Factors of Acceptance

1. The need for change
2. Scope for greater satisfaction
3. Utility has been proved
4. Compatible with existing culture
5. Cost for accepting change
6. Role of communicator

The communication behaviour of rural women is the sum total of expression ability, credibility, characteristics of message, treatment of message and use of appropriate channel. The scores obtained on each account can be calculated and the level of communication can be classified as high, medium and low.

Measurement of Innovativeness of Rural Women

Innovativeness is the degree to which an individual is relatively earlier in a social system. It is the degree to which an individual is sensitive to adopt an innovation. With advancement in science and technology, there is constant evolution of innovations to manage. Women when sensitive to adoption of new practices there occurs a series of changes in attitude formation. When positive attitude dominates over negative ones innovations are adopted. Innovations are observed on every walk of life. Women consider innovation more seriously and look them from all people practical points of view before adoption than men. Measurement of innovativeness is relatively difficult. It needs time factor for examination. The total number of innovations adopted by person within a time frame helps to decide innovativeness.

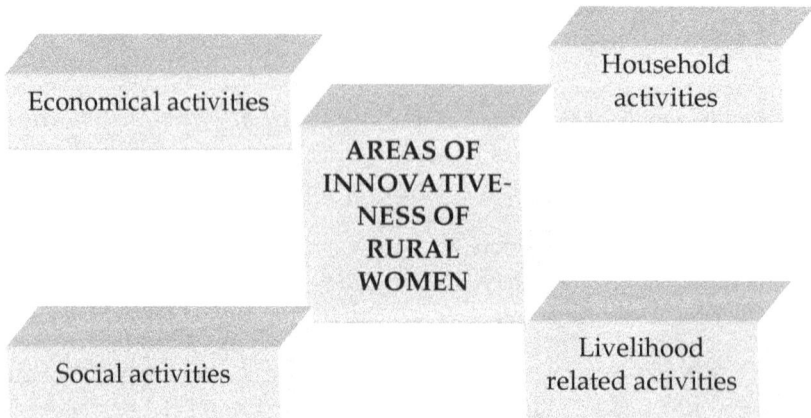

Economical activities

Household activities

AREAS OF INNOVATIVE-NESS OF RURAL WOMEN

Social activities

Livelihood related activities

The operational definition of innovativeness measurement is the number of innovations adopted by women within specific period of time. A three point continuum may be used to measure the innovativeness in different areas of women concern.

Sl.No.	Activities	Extent of Adoption		Time Taken in Years
		Full	Partial	
1.	**Family activities**			
a.	Cooking methods			
b.	New food items			
c.	Child care			
d.	Sanitary practices			
e.	House hold planning			
2.	**Economic Activities**			
a.	Money saving			
b.	Judiciuos expenditure			
c.	Balanced investment			
d.	Purchase of assets			
e.	Disposal of produce			
f.	Economic use of resources			
g.	Burrowing/lending money			
h.	Economic planning			
i.	Maintaining economic status			
3.	**Social activities**			
a.	Social relationship			
b.	Participation in social functions			
c.	Involvement in community work			
d.	Maintaining cohesiveness in family			
e.	Maintaining of social status			
4.	**Livelihood activity**			
a.	Adoption of new crop/varieties			
b.	Use of optimum inputs			
c.	Maintaining of livestock			
d.	Use of labour			
e.	Operation of non-farm unit			

We can assign 2 and 1 score for full and partial adoption of the innovation per time unit. There may be two way classifications.

1. **Time Factor**: Taking time as basic factor we can classify the innovativeness.

Sl.No.	Time Period	Level of Innovativeness
1.	Less than one year	High
2.	Two to three years	Medium
3.	More than three years	Low

2. On the basis of score, maximum obtainable score = 50 minimum score = 25

Sl.No.	Category	Score
1.	No innovativeness	up to 5
2.	Low	6-17
3.	Medium	18-32
4.	High	33 and above

On the basis of innovativeness, the adopter categories are classified as:

(i) Innovators   (2.5 per cent)

(ii) Early adopters   (13.5 per cent)

(iii) Early majority (34 per cent)

(iv) Late majority   (34 per cent)

(v) Laggards   (16 per cent)

Such classification does not hold good so far as our social system and living pattern of rural women in Odisha is concerned.

Communication Index

The communication index can be calculated adopting judge rating methods and assigning values to different factors out of 100 as felt appropriate by the judges.

Communication Index

Factors	J1	J2	J3	J4	J5	J6	J7	J8	J9	J10	Average
1. Expression ability	20	12	11	10	13	15	18	8	12	14	13.3
2. Credibility	25	18	30	25	20	28	25	21	27	31	25.0
3. Message	30	36	33	35	32	22	16	39	20	19	28.2
4. Message treatment	15	18	12	20	23	22	32	14	24	21	19.1
5. Use of channel	10	16	14	10	12	13	9	18	17	15	13.4
TOTAL	100	100	100	100	100	100	100	100	100	100	19.8

Sl.No.	Factors	Weight Age
1.	Expression ability	13.30
2.	Credibility	25.00
3.	Message	28.20
4.	Treatment of message	19.10
5.	Channel	13.40

The index value of each factor of communication may be used for further statistical analysis.

Chapter 7
Measurement of Managerial Ability of Rural Women

Women are the managers of individual families. They play important role in family management. The good management of family depends upon managerial ability of the housewives with support of the family members. The good and smooth management leads to development and prosperity of family as a whole. Measurement of managerial ability could provide a clear picture about efficiency of family managers and deficiencies if any. The major activities of rural housewives are:

1. *Household activities*: Kitchen management, food, nutrition, health care, sanitation, drinking water
2. *Care of children and their education*: Child care, schooling, progress in study, dress and personality development
3. *Resource management*: Components of livelihood system, crop, livestock, land improvement, non-farm activities, investment/expenditure
4. *Social functions and community activities*: Social functions, village committee, development work, Self Help Group, common property management.

These are the activities in which the managerial ability of women play important role. The managerial ability can be measured to decide basis for capacity building. A four point continuum of very much, much, some extent and never could be used to measure managerial ability of women with assigned score of 4, 3, 2 and 1 respectively.

1. Household Activities

A simple measurement of family life on the parameters of smooth functioning

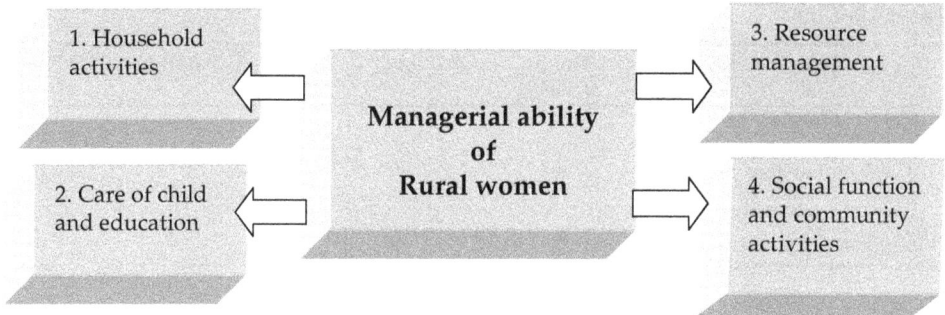

Figure 17: Dimensions of Managerial Ability of Rural Women.

and presence of happiness reveals the managerial ability of housewives. The format given below can be used to measure the managerial ability in household activities.

Very much = 4, Much = 3, Some extent = 2 and Not at all = 1 (put the number only as you decide)

Table 42: Managerial Ability of Rural Women on Household Activities

Managerial Ability	Kitchen Nutrition	Food and	Health	Sanitation	Drinking Water
1. Planning					
a. Advance planning					
b. Consideration of past experience					
c. Discussion with others					
2. Organizing					
a. Fixing priority					
b. Action on priority					
c. Organization of material					
3. Supervising					
a. Supervise work					
b. Stress on completion in time					
4. Communication					
a. Tell about task in time					
b. Discuss about work					
c. Listen to others					
d. Gather more information					
5. Coordinating					
a. Arrangement of materials before starting					
b. Understanding with all					
c. Looking for cooperation					

Contd...

Table 42–*Contd...*

Managerial Ability	Kitchen Nutrition	Food and	Health	Sanitation	Drinking Water
6. Controlling					
a. Taking remedial step					
b. Ensure alternative					
c. Control over expenditure					

Maximum Score = 4X (18 × 5) = 360 Minimum score = 1X (18 × 5) = 90

Sl.No.	Categorization of Managerial Ability	Score Range
1.	High	268 and above
2.	Medium	180-267
3.	Low	up to 179

1. Planning
2. Organizing
3. Supervising
4. Communication
5. Coordinating
6. Controlling

Besides the categorization, the gap in managerial ability can also be calculated by finding out difference between achievable scores and scores obtained by the respondent in percentage.

2. Care of Children and their Education

Universally mothers play a very important role in child care and their education. Now there is remarkable change in rural areas concerning to the education of children. The mothers are now more conscious than earlier days. The managerial ability of rural women is reflected in educational attainment of their children. How much managerial abilities our rural women possess can be measured in a three point continuum assigning 3, 2 and 1 for always, sometimes and never respectively.

Question

Care of children and their education involve health, admission in school, study progress, personality development, etc. To what extent do you perform the following responsibilities and how do you follow the following guidelines of management (N = 50, Jatni block, 20012).

Table 43: Managerial Ability of Rural Women on Child Care and Education

Managerial Ability	Always	Sometimes	Never	Average	Gap (per cent)
1. Planning					
a. Advance planning	12	18	20	1.84	38.66
b. Consideration of past experience	10	15	25	1.70	43.33
c. Discussion with others	8	12	30	1.56	48.00

Contd...

Table 43–*Contd...*

Managerial Ability	Always	Sometimes	Never	Average	Gap (per cent)
2. Organizing					
a. Fixing priority	18	10	22	1.88	37.33
b. Action on priority	10	20	20	1.80	40.00
c. Organization of study material	9	12	29	1.60	46.66
3. Supervising					
a. Supervise progress of children education	16	14	20	1.92	36.00
b. Stress on completion in time	7	12	31	1.52	49.33
4. Communication					
a. Tell about task in time	8	18	24	1.68	44.00
b. Discuss about study	15	15	20	2.50	16.66
c. Listen to problems of children	30	15	5	2.50	16.66
d. Collecting information about study	6	12	32	1.48	50.66
5. Coordinating					
a. Arrangement of study materials before starting	13	13	24	1.78	40.66
b. Contact with teachers	10	20	20	1.80	40.00
6. Controlling					
a. Taking remedial step when children fail in examination	14	14	22	1.84	38.66
b. Restrict non-school activities	11	16	23	1.76	41.33
c. Control expenditure	13	17	20	1.86	38.00
Average	—	—	—	1.82	39,33

The overall managerial ability gap is 39.33 per cent so far as child care and their education is concerned. Taking other rout, the classification on managerial ability, the sample may be grouped as high, medium and low.

Maximum obtainable score = $17 \times 3 = 51$

Minimum obtainable score = 17

Sl.No.	Categorization of Managerial Ability	Score Range
1.	High	39 and above
2.	Medium	29-38
3.	Low	28 and less

The managerial ability indicated through gap analysis will form basis for capacity building programme. The sample women are deficient of managerial ability to look after education of children effectively. The managerial ability of rural women depends on various factors.

1. Education of self
2. Education of parent
3. Environment of past and present
4. Amount of freedom availed
5. Exposure to outside world
6. Cooperation of partner and family member
7. Community influence
8. Exposure to media
9. Socio-economic status of the family
10. Age and experience

Experience is the best teacher in life

(The ten variables are measurable and have been measured by many scholars)

3. Resource Management

Resource management in rural areas is a multidimensional work. It needs group approach. Resources in family are mostly farm based and non-farm based. The components of livelihood system are: crop, livestock, land improvement, non-farm activity, labour force, investment/expenditure, etc. Both male and female manage these resources. The management ability of either sex counts much for success. The measurement of managerial ability of women in managing the resources can be performed applying basic rules of management.

Question

To manage crop, live stock, land, irrigation, labour force and non-farm activities to what extent you follow the following steps (N = 50, Jatni Block, 2012).

Managing of crop, live stock, land, irrigation, labor force and non-farm activities etc. (Full extent = 3, Partially = 2, Not at all = 1)

Table 44: Managerial Ability of Farm Women on Resource Management

Managerial Ability	Full Extent	Partially	Not at All	Mean Score	Gap (per cent)
1. Planning					
a. Advance planning	21	13	16	2.10	30.00
b. Consideration of past experience	17	12	21	1.92	36.00
c. Discussion with others	8	12	30	1.56	48.00
2. Organizing					
a. Fixing priority	15	13	22	1.86	38.00
b. Action on priority	15	18	17	1.96	34.66
c. Organization of materials before start	9	17	24	1.70	43.33

Contd...

Table 44–*Contd...*

Managerial Ability	Full Extent	Partially	Not at All	Mean Score	Gap (per cent)
3. Supervising					
a. Supervise work	8	11	31	1.54	48.66
b. Stress on completion in time	10	11	29	1.62	46.00
4. Communication					
a. Tell about task in time	6	9	35	1.42	52.66
b. Discuss about work	7	12	31	1.52	49.33
c. Listen to others	7	8	35	1.44	52.00
d. Gather more information	4	8	38	1.32	56.00
5. Coordinating					
a. Procurement of materials before starting	13	12	25	1.76	41.33
b. Understanding with all about resources	3	6	41	1.24	58.66
c. Looking for cooperation	5	8	38	1.38	54.00
6. Controlling					
a. Taking remedial step	10	12	28	1.64	45.33
b. Pointing out deviations	5	4	41	1.28	57.33
c. Control over expenditure	16	17	17	1.98	34.00
Average	-	-	-	1.53	49.00

The maximum score = 54, minimum score = 18. The classification of respondents on managerial ability in resource management may be high medium and low.

Sl.No.	Category	Range of Score
1.	High	42 and above
2.	Medium	31-41
3.	Low	30 and less

The gap analysis will also provide useful information and can be considered as basis to formulate programme for capacity building.

Sl.No.	Management Parameters	Average Score	Gap (per cent)
1.	Planning	1.86	38.00
2.	Organizing	1.61	46.33
3.	Supervising	1.58	47.33
4.	Communication	1.42	52.66
5.	Coordination	1.46	51.33
6.	Controlling	1.63	45.66

4. Management of Social Functions and Community Activities

At present importance of rural women is gaining momentum in rural development activities. Their participation in social functions and community activities is increasing day by day. These activities are, social functions like marriage, death ceremonies, village cultural functions, village committee, development work, Self Help Group and common property management. To manage all these activities, management efficiency is a must. To what extent our rural women are capable of performing these activities depend on their managerial abilities. The managerial ability can be measured following the procedures discussed above.

On the whole, the managerial abilities can be computed taking calculated values on account of: (i) household activities, (ii) child care and their education, (iii) resources management and (iv) socio-community activities.

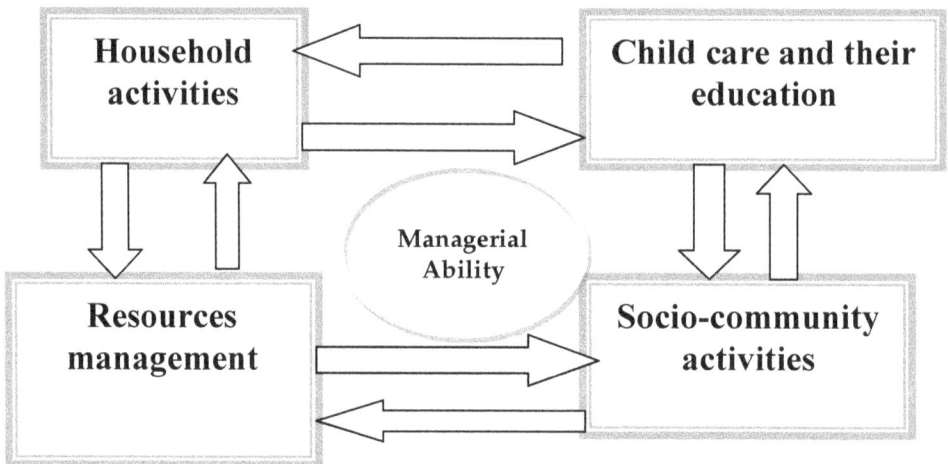

Figure 18: Managerial Ability.

Management Traits of Rural Women

Women constitute around half of the total world population. This also holds well in India. They are now participating outside home activities. The woman entrepreneurship now is in rising trend. To be in this field they need to possess managerial traits. Women entrepreneurs may be defined as a woman or group of women, who initiate, organize and run business enterprise. The women who innovate, imitate or adopt a business activity also come under this category. The Government of India has defined women entrepreneurs based on women participation in equity and employment of a business enterprise. According to Khanka (1999) women entrepreneurs are those women who think of business enterprise, initiate it, organize and combine the factors of production, operate the enterprise and undertake risk and handle economic uncertainty involved in running a business enterprise. The definition holds good for rural women who are involved in management of livelihood irrespective of size and volume. The functions relating to family management involves: (i) risk taking, (ii) organizing inputs and (iii) adding of innovations for improvement of living condition. The management of any enterprise or family requires the traits like:

1. Hard work
2. Desire for achievement
3. Optimism
4. Independence
5. Foresight
6. Organizing capacity
7. Innovativeness
8. Risk bearing attitude

These traits are observed with women of rural areas may be in differential proportion. These traits are to be induced in them to increase their efficiency. A study conducted in Odisha taking members of women dairy cooperative societies adopted three point continuum to measure these traits.

Sample: Members of women dairy

Co-operative societies

Sample size= 110

Location: Dhenkanal and Cuttack districts of Odisha

Measurement: Three point continuum, very much, much and little (3, 2 and 1)

Analysis: Score analysis and gap determination (per cent).

Table 45: Managerial Traits of Rural Women

Sl.No.	Managerial Traits	Cuttack	Dhenkanal	Average	Gap (per cent)	
1.	Hard work	1.84	1.18	1.51	50.66	
2.	Desire for achievement	2.04	1.57	1.80	40.00	
3.	Optimistic	2.34	2.03	2.18	27.33	
4.	Foresightness	1,70	1.11	1.40	62.00	Maximum Score = 30
5.	Independence	1.43	1.38	1.40	53.33	
6.	Planning ability	1.45	1.26	1.35	55.00	Minimum Score = 10
7.	Organizing ability	1.60	1.56	1.58	47.33	
8.	Innovativeness	1.70	1.61	1.65	45.00	
9.	Information seeking	1.60	1.23	1.41	53.00	
10.	Quality consciousness	1.45	1.o4	1.24	58.66	
	Average	1.71	1.39	1.55	49.73	

Swain, P, 2006 Performance of women dairy cooperative societies in Orissa: a critical analysis, Ph.D. Thesis, OUAT, Bhubaneswar, p.178.

Women entrepreneurs also face the problems. The problems are: (i) problem of finance as the properties are not in their names, (ii) scarcity of raw materials in many cases as they are not able to organize, (iii) stiff competition in the market, (iv) limited mobility, (v) family ties, (vi) lack of education, (vii) dominancy of male members and (viii) low risk bearing capacity. These problems are measurable provided their intensity is observed at a significant level.

Entrepreneurship Development Institute of India (EDI) in its study to identify a set of entrepreneurial competencies or characteristics have noted the following indicators.

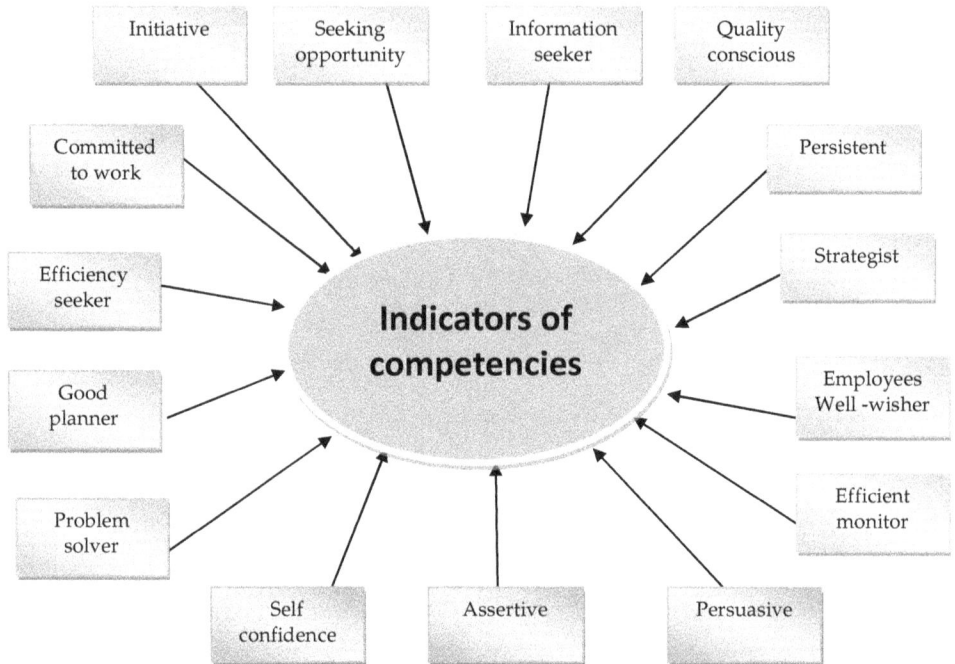

Figure 19: Indicators of Competencies

(All these indicators can be very well measured)

Concept of Best Management Practices

At present our attention is on good management practices to derive maximum benefit. The concept is more familiar in agricultural practices. The concept emphasizes three important aspects. These are: (i) effective use of resources, (ii) maximize benefit from export agro practices and (iii) use of sustainable agriculture.

In case of rural women this concept has much applicability. The best management practices should have five elements. Elements of Best Management Practices as applicable in case of rural women are:

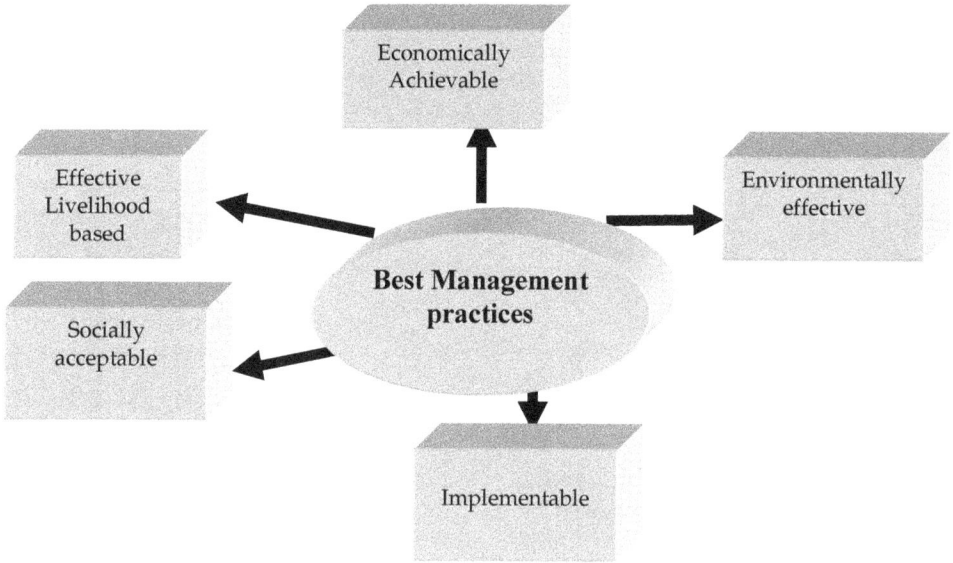

Figure 20: Management Practices.

How to Measure these Elements?

Human Capital

1. *Effective livelihood base*: Livelihood base of each and every household in villages are almost same with variation in size. The sustainable livelihood frame work can be used as analytical tool to identify and assess internal and external factors to the households that affects its socio-economic survival. The capital assets pentagon of sustainable livelihood system (Chambers and Conway, 1992) have presented as follows:

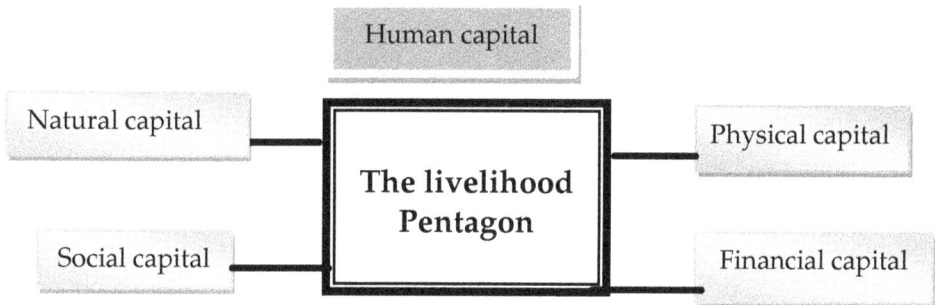

Figure 21: The Livelihood Pentagon.

1. Social capital, 2. Financial capital, 3. Natural resources, 4. Physical capital and 5. Human capital

All these capitals are measurable in quantitative and qualitative terms. For qualitative measurement a scale has to be developed looking into location of study and three point continuums can be used to measure them. Search for Sustainable Livelihood System: Managing resources and Change (Baumgartner and Hogger, 2004) explained the livelihood system in using Roof Concept in term of *mandals*.

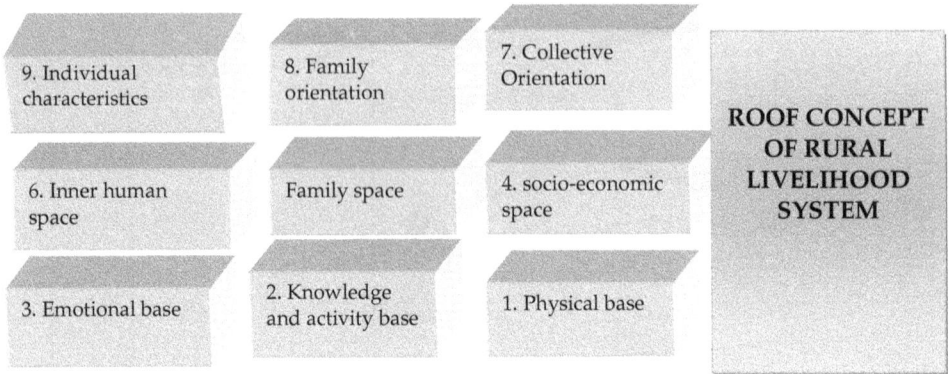

Figure 22: Concept of Rural Livelihood System.

Nine variables shown in nine cells can be measured in qualitative term.

Table 46: Livelihood Concept and Components

Sl.No.	Author(s)	Concepts
1.	Chambers 1989	Different sources, adjustment measures, institutional mechanism, production, investment
2	Asian Development Bank 1991	Sustainability, ecosystem, overtime, non-declining output despite shocks
3.	Robert Chambers & Conway 1992	Comprises the capabilities, assets, stores, resources, claims and access, activities, opportunities for next generation, contribute to other livelihood system
4.	CARE 1994	Capacity building and passive external help
5.	Lucas 1997	Livelihood capital in terms of human capital consisting of education, skill, knowledge and health, access to economic opportunities
6.	UNDP 1998	Comprises capabilities, assets and activities required for a means of living
7.	**Scoones ?**	Stated of four types of capital *i.e.*, natural, financial, human, and social capital
8.	Farrington and James 2000	Watershed, diversification forestry, pasture development, livestock development, micro enterprise development
9.	Conroy *et al.*, 2001	Marginal, small farmers, livestock, daily wages in small and marginal household
10.	DFID 2005	Capabilities, assets, activities required for living
11.	Mearns 2005	Micro-level institutional context, status and deprivation

References

1. Baumgartner, R. and Hogger, R. (2004) *'Search for Sustainable Livelihood System'*, SAGE Publication, New Delhi, India.

2. Economic Survey (2008-09) Government of India, Department of Economic Affairs, Oxford University Press, New Delhi, India.

3. Edwards, A. L. (1969) *'Techniques of Attitude Scale Construction'*, Vakils, Feffer and Simons Pvt. Ltd, Bombay, India.

4. Kaushik, U. and Bhatnagar, S. (2007) *'Entrepreneurship'*, Aavishkar Publisher Distributors, Jaipur, India.

5. Khanka, S. S. (2004) *'Entrepreneurial Development'*, S. Chand and Companies Ltd, New Delhi, India.

6. Malkal, K. and Reddy, P. R. (2007) *'Gender Issues'*, Haritha Publishing House, Secunderabad, India.

7. Northouse, P. G. (2007) *'Leadership'*, SAGE Publication, New Delhi, India.

8. Roy, R. (2011) *'Entrepreneurship'*, Oxford University Press, New Delhi, India.

9. Sah, A. K. (1992) *'Systems Approach to Training and Development'*, Sterling Publishers Pvt. Ltd, New Delhi, India.

10. Satapathy, C. and Mishra, S. (2008) *'Extension Techniques for Rural Management'*, Kalyani Publisher, Ludhiana, India.

Index

A

Accurate 69
Adoption 64
Agriculture engineering 47
Agronomy 47
AIDS 1
Articulation 31
Assessment 50
Awareness indicators 25

B

Before-after design 53

C

Calorie 18
Capacity building 42
Categorization 70
Channel selection 71
Child related decision 22
Children 78
Communication 67
Communication behaviour 65
Communication index 74

Community activities 82
Components 3
Convincing 69
Credibility 67
Cultural factors 64

D

Decision-making 2, 6, 21
Dependent variables 16
Determination 29
Determined 31
Development 5, 7
Development variables 15
Discrimination 14
Drudgery 14

E

Economic empowerment 24
Economic factors 64
Economic front 7
Economic productivity 12
Economic status 11
Economic variables 11

Education 3, 5, 17, 78
Employment 17, 20
Empowerment 2, 3, 8
Equity 7
Experimental research design 52
Expression ability 67
Extension contact 15
Extension education 46
Extension factors 64
Extension participation 15
External variable 55

F

Family income 12
Farm related decision 22
Farm size 11
Farm women 56, 65
Financial decision 22
Financial management 5
Fishery 46

G

Gap analysis 70
Gender development index 4, 26
Gender empowerment 4, 7
Gender equity 16
Gender equity index 4, 20

H

Health 3
Health and nutrition 18
Home science 48
Horticulture 44
Household activities 76
Household exchange 23
Human capital 85
Human skill 32

I

Income generating capacity 12
Independent variables 10

Indicators 9
Innovativeness 56, 72
Integrity 29
Intelligence 29
International development 1
Interrelationship 9

K

Knowledge Gain 50

L

Labour allocation 22
Leadership 37
Legal empowerment 26
Livestock 19
Livestock related decision 22

M

Managerial ability 13, 76
Marketing factors 64
Mass media exposure 15
Measure adoption behaviour 57
Measurement 7, 8, 28, 34, 50, 55
Message 69
Millennium development goal 1
Minimization 55

N

NGO 43
Nutritional status 18

O

OXFAM 43

P

Plant protection 45
Political awareness 24
Political empowerment 24
Political level 8
Political participation 24
Productivity gain 6
Protein 18

Psychological level 8
Psychological traits 5
Psychological variables 4, 14

R

Resource management 80
Rural women 2, 72, 76, 82

S

Self confidence 29
Self orientation 14
SHG 3, 39
Situational approach 35
Skill approach 32
Sociability 29
Social empowerment 23
Social mobilization 11
Social variables 10
Socio-cultural level 7
Socio-religious obligation 23
Style approach 33

T

Technical skill 32
Technological factors 64
Training 43
Training programme 49
Trait approach 29
Traits 28
Trustworthy 31

U

UNDP 2, 43
UNICEF 2

V

Variables 9

W

Women 5
World customs organization 43
World development report 5

www.ingramcontent.com/pod-product-compliance
Lightning Source LLC
Chambersburg PA
CBHW020753300326
41914CB00050B/184